Lecture Notes
in Business Information Processing   28

Series Editors

Wil van der Aalst
Eindhoven Technical University, The Netherlands
John Mylopoulos
University of Trento, Italy
Norman M. Sadeh
Carnegie Mellon University, Pittsburgh, PA, USA
Michael J. Shaw
University of Illinois, Urbana-Champaign, IL, USA
Clemens Szyperski
Microsoft Research, Redmond, WA, USA

Erik Proper   Frank Harmsen
Jan L.G. Dietz (Eds.)

Advances in Enterprise Engineering II

First NAF Academy Working Conference
on Practice-Driven Research on Enterprise Transformation
PRET 2009, held at CAiSE 2009
Amsterdam, The Netherlands, June 11, 2009
Proceedings

 Springer

Volume Editors

Erik Proper
Capgemini and Radboud University Nijmegen
Toernooiveld 1, 6525 ED, Nijmegen, The Netherlands
E-mail: e.proper@acm.org

Frank Harmsen
Capgemini and University of Maastricht
Minderbroedersberg 4-6, 6211 LK Maastricht, The Netherlands
E-mail: frank.harmsen@capgemini.com

Jan L.G. Dietz
Delft University of Technology
Mekelweg 4, 2628 CD Delft, The Netherlands
E-mail: j.l.g.dietz@tudelft.nl

Library of Congress Control Number: Applied for

ACM Computing Classification (1998): J.1, H.4, D.2

ISSN   1865-1348

ISBN 978-3-642-01858-9 Springer Berlin Heidelberg New York

springer.com

© Springer-Verlag Berlin Heidelberg 2009

Typesetting: Camera-ready by author, data conversion by Scientific Publishing Services, Chennai, India
Printed on acid-free paper   SPIN: 12681907   06/3180   5 4 3 2 1 0

Advances in Enterprise Engineering

Enterprise engineering is an emerging discipline that studies enterprises from an engineering perspective. Two key paradigms underpin this discipline. The first paradigm states that enterprises are purposefully designed and implemented systems. Consequently, they can be re-designed and re-implemented if there is a need for change. All kinds of changes are accommodated: strategic, tactical, operational, and technological. The second paradigm of enterprise engineering is that enterprises are social systems. This means that the system elements are social individuals, and that the essence of an enterprise's operation lies in the entering into and complying with commitments between these social individuals.

Enterprise engineering is rooted in both the organizational sciences and the information system sciences. In our current understanding, three concepts are paramount to the theoretical and practical pursuit of enterprise engineering: enterprise ontology, enterprise architecture and enterprise governance. Enterprise ontology concerns the understanding of an enterprise in a way that is fully independent of any implementation. The ontological model of an enterprise shows the essence of its operation. It is the starting point for designing and implementing all kinds of changes. It is also extremely stable over time; most changes appear to be changes in the implementation. Enterprise architecture concerns the identification, the specification, and the application of design restrictions, which come in addition to the specific requirements in every change project. These design restrictions provide an operationalization of an enterprise's strategic basis (mission, vision), and offers restrictions and guidance on how to shape and implement the ontological model of the enterprise. Only in this way can one achieve and guarantee that the operations of an enterprise are fully compliant with its mission and strategies. Lastly, enterprise governance constitutes the organizational conditions for incorporating enterprise ontology and enterprise architecture in an enterprises practice. It constitutes the primary condition for making the enterprise engineering approach feasible and beneficial.

The vast majority of strategic initiatives fail, meaning that enterprises are unable to gain success from their strategy. The high failure rates are reported from various domains: total quality management, business process reengineering, six sigma, lean production, e-business, customer relationship management, as well as from mergers and acquisitions. It appears that these failures are mostly the avoidable result of an inadequate implementation of the strategy. Rarely are they the inevitable consequence of a poor strategy. Abundant research indicates that the key reason for strategic failures is the lack of coherence and consistency, collectively also called congruence, among the various components of an enterprise. At the same time, the need to operate as an integrated whole is becoming increasingly important. Globalization, the removal of trade barriers, deregulation, etc., have led to networks of cooperating enterprises on a large scale, enabled by the virtually unlimited possibilities of modern information and communication technology. Future enterprises will therefore have to operate in an ever more dynamic and global environment. They need to be more agile, more adaptive, and more transparent. In addition, they will be held more publicly accountable for every effect they produce.

These challenges are traditionally addressed by black-box thinking-based knowledge, i.e., knowledge concerning the function and the behavior of enterprises, as contained in the organizational sciences. Such knowledge is su cient, and perfectly adequate, for managing an enterprise (within the range of control). However, it is definitely inadequate for changing an enterprise. In order to bring about changes, white-box-based knowledge is needed, i.e., knowledge concerning the construction and the operation of enterprises. Developing and applying such knowledge requires no less than a paradigm shift in our thinking about enterprises, since the organizational sciences are dominantly oriented toward organizational behavior, based on black-box thinking.

The current situation in the organizational sciences resembles very much the one that existed in the information system sciences around 1970. At that time, a revolution took place in the way people conceived information technology and its applications. Since then, people have been aware of the distinction between the form and the content of information. This revolution marks the transition from the era of data systems engineering to the era of information systems engineering. The comparison we draw with the information system sciences is not an arbitrary one. On the one hand, the key enabling technology for shaping future enterprises is the modern information and communication technology (ICT). On the other hand, there is a growing insight into the information system sciences that the central notion for understanding profoundly the relationship between organization and ICT is the entering into and complying with commitments between social individuals. These commitments are raised in communication, through the so-called intention of communicative acts. Examples of intentions are requesting, promising, stating, and accepting. Therefore, as the content of communication was put on top of its form in the 1970s, the intention of communication is now put on top of its content. It explains and clarifies the organizational notions of collaboration and cooperation, as well as authority and responsibility. It also puts organizations definitely in the category of social systems, very distinct from information systems. Said revolution in the information systems sciences marks the transition from the era of information systems engineering to the era of enterprise engineering, while at the same time merging with relevant parts of the organizational sciences, as illustrated in the figure below.

The mission of the discipline of enterprise engineering is to combine (relevant parts from) the organizational sciences and the information system sciences, and to develop theories and methodologies for the analysis, design, and implementation of future enterprises. Two crucial concepts have already emerged that are considered paramount for accomplishing this mission: enterprise ontology and enterprise architecture. A precondition for incorporating these methodologies effectively in an enterprise is the good establishment of enterprise governance.

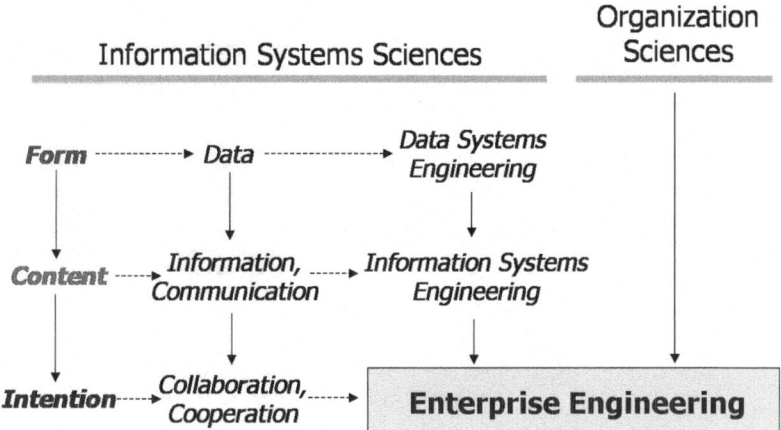

To establish and further develop the discipline of enterprise engineering, a Springer series on Enterprise Engineering has been established. Books in this series are aimed at academic students and advanced professionals, while their content ranges from theoretical foundations to application experiences. The *Advances in Enterprise Engineering* have been created, within LNBIP, to stimulate active research in this field. They are dedicated to proceedings of conferences and workshops aiming to explore the boundaries of the field of enterprise engineering, to deepen the understanding of the field, as well as to study its application in practice.

We are glad to welcome the First NAF Academy Working Conference on Practice-Driven Research on Enterprise Transformation to the *Advances in Enterprise Engineering*. Thematically, this conference takes a wide perspective on enterprise transformation while acknowledging that as a discipline enterprise engineering is at the heart of succesful enterprise transformations. At the same time, studying enterprise transformation with a wide perspective also enables the further operationalization of the requirements and challenges to be met by the enterprise engineering discipline. Furthermore, this working conference focuses explicitly on research results bridging academia and industry. In an emerging field such as enterprise engineering it is of the utmost importance to study challenges from industrial practice as well as the applicability of research results in practice.

March 2009

Jan L.G. Dietz
Erik Proper
Frank Harmsen

Practice-Driven Research on Enterprise Transformation

Modern-day enterprises, be they businesses or organizations, are in a constant state of flux. New technologies, new markets, globalisation, mergers, acquisitions, etc. are among the "usual suspects" which require enterprises to transform themselves to deal with these challenges and new realities. Most information systems practitioners will find themselves working in a context of enterprise transformation. One could even go as far as to claim that a business-oriented perspective on the evolution of information systems is really about enterprise transformation, where enterprise transformation involves the use of methods and techniques from enterprise engineering, enterprise modeling, enterprise architecture, and information systems engineering.

As a field of study, enterprise transformation requires a close interaction between practice and academia. What works and does not work requires validation in real-life situations. Conversely, it is in industrial practice where challenges can be found that may fuel and inspire researchers. This sparked the idea to create an industrial track on "Practice-Driven Research on Enterprise Transformation" at CAiSE 2009. From the start, however, the intention was to run such an event more often and let it become more than a one-off event.

The Netherlands Architecture Forum (NAF) is a Netherlands-based organization fostering the development of the field of IT and enterprise architecture as a means to aid in the informed governance of transformations of enterprises and their IT. NAF is an organization of businesses and organizations, and has well over 70 members, covering three domains:

1. Universities, universities of applied science, and research institutes
2. Organizations providing architecture-related services
3. Organizations using architecture-related services

It is the desire of NAF to stimulate interaction between industry and academia in the area of architecture, while acknowledging that architecture is a means to aid in the informed governance of enterprise transformations. This led the organizers of the industrial track to combine forces with NAF, to create the *NAF Academy* as a label under which to continue organizing the "Practice-Driven Research on Enterprise Transformation" industrial track as an *annual working conference*. This working conference should be attractive to visitors from both academia and industry, and as such aims to create an open environment in which to stimulate the knowledge exchange between both worlds. To further stimulate this exchange, we aim to continue organizing the *NAF Academy Working Conference on Practice-Driven Research on Enterprise Transformation* in co-location with other relevant scientific events.

The proceedings of the NAF Academy working conference will be organized in terms of a small number of longer papers of about 20 pages on average, while the program of the actual event will focus on interaction between theory and practice. In this inaugural year, we already received 30 high-quality submissions. From these submissions the Program Committee selected 11 submissions based on their scientific quality as well as their potential in bridging the gap between industry and academia.

March 2009   Erik Proper

Organization

Steering Committee

Frank Harmsen   Capgemini and University of Maastricht,
     The Netherlands
Erik Proper   Capgemini and Radboud University Nijmegen,
     The Netherlands

Organizing Committee and Program Board

Chair: Erik Proper   Capgemini and Radboud University Nijmegen,
     The Netherlands
Frank Harmsen   Capgemini and University of Maastricht,
     The Netherlands
Huub Bakker   Atos Origin, The Netherlands
Egon Berghout   M&I/Partners and University of Groningen,
     The Netherlands

Denis Hageman   Yacht, The Netherlands
Nico Lassing   Accenture, The Netherlands
Bas van der Raadt   Ernst & Young, The Netherlands
Frank Schalkwijk   Atos Origin, The Netherlands
Zinze Siegerink   Capgemini, The Netherlands

Program Committee

Anne Persson   University of Skövde, Sweden
Bas van Gils   Strategyworks, The Netherlands
Bas van der Raadt   Ernst & Young, The Netherlands
Camille Salinesi   University of Paris 1, France
Denis Verhoef   Kirkman Company, The Netherlands
Egon Berghout   University of Groningen, The Netherlands
Erik Proper   Capgemini and Radboud University Nijmegen,
     The Netherlands
Frank Harmsen   Capgemini and University of Maastricht,
     The Netherlands
Geert-Jan Houben   Delft University of Technology,
     The Netherlands
Gregor Engels   SDM and University of Paderborn, Germany
Hajo Reijers   Eindhoven University of Technology,
     The Netherlands

Hans Mulder	VIA Groep, The Netherlands and University of Antwerp, Belgium
Henry Franken	BizzDesign, The Netherlands
Jaap Gordijn	VU University Amsterdam, The Netherlands
Jan Dietz	Delft University of Technology, The Netherlands
Jan Hoogervorst	Sogeti, The Netherlands
Jan Mendling	Queensland University of Technology, Australia
Johan Versendaal	University of Utrecht, The Netherlands
Marc Lankhorst	Telematica Instituut, The Netherlands
Marta Indulska	University of Queensland, Australia
Nico Lassing	Accenture, The Netherlands
Patricia Lago	VU University Amsterdam, The Netherlands
Pnina Soffer	University of Haifa, Israel
Raymond Slot	Capgemini, The Netherlands
Robert Winter	University of St. Gallen, Switzerland
Sadie Legard	Corus, UK
Stefan Strecker	Duisburg-Essen University, Germany
Stijn Hoppenbrouwers	Radboud University Nijmegen, The Netherlands
Wolfgang Hesse	Philipps University Marburg, Germany

Endorsing Organizations

Conference on Advanced Information Systems Engineering (CAiSE)
IFIP Working Group 8.1
The Benelux Chapter of the Association for Information Systems (AIS), The Netherlands
The Netherlands Architecture Forum (NAF), The Netherlands
The GI-Fachgruppe MobIS of the German Computer Society, Germany
Via Nova Architectura, The Netherlands

Supporting Organizations

Accenture, The Netherlands
Atos Origin, The Netherlands
Capgemini, The Netherlands
M&I/Partners, The Netherlands
The Netherlands Architecture Forum (NAF), The Netherlands
University of Groningen, The Netherlands
Radboud University Nijmegen, The Netherlands

Table of Contents

A Holistic Software Engineering Method for Service-Oriented Application Landscape Development

Andrea Baumann[1], Gregor Engels[1,2], Alexander Hofmann[1], Stefan Sauer[2], and Johannes Willkomm[1]

[1] Capgemini sd&m AG, Carl-Wery-Str. 42, 81739 Munich, Germany
{andrea.baumann,alexander.hofmann,
johannes.willkomm}@capgemini-sdm.com
[2] s-lab – Software Quality Lab,
University of Paderborn,Warburger Str. 100, 33098
Paderborn, Germany
{engels,sauer}@s-lab.upb.de

Abstract. Enterprises are transforming into enterprises which follow from a business as well as from an IT perspective a service-oriented paradigm. This change towards service-oriented enterprise and IT architectures has to be reflected in the methodologies of developing whole application landscapes as well as individual applications. Quasar (Quality Software Architecture) has been developed as the standard architecture and development method of Capgemini sd&m for individual applications. For the development of service-oriented enterprise application landscapes, Quasar Enterprise has been designed. Both Quasar and Quasar Enterprise are integrated with each other within a holistic software engineering method to seamlessly cover the full development lifecycle of service-oriented application landscapes, from business modeling and service design to actual software development. In this paper, we illustrate how a company-wide ontology of development artifacts serves as the key feature for integrating both methods.

Keywords: Service-oriented Architecture, Managing/Governing Enterprise Evolution, Business-IT Alignment, Software Development Techniques.

1 Introduction

Modern day enterprises are transforming into enterprises which follow from a business as well as from an IT perspective a service-oriented paradigm. This allows for alignment of enterprise business processes and application landscapes, which consist of interacting IT applications. By this, high flexibility and adaptability in case of changing business requirements or changing IT technologies is facilitated.

As a direct consequence, this change of the underlying paradigm towards designing and implementing service-oriented enterprise and IT architectures has to be reflected in the methodologies of developing whole application landscapes as well as individual applications. Therefore, software companies who develop enterprise applications have

E. Proper, F. Harmsen, and J.L.G. Dietz (Eds.): PRET 2009, LNBIP 28, pp. 1–17, 2009.

to be transformed, too. They do no longer focus solely on small- or medium-sized applications, but target at large-scale application landscapes.

This is particularly true for the software company Capgemini sd&m which is successfully active in the market for more than 25 years. Within numerous projects, custom-made, domain-specific applications (predominantly business and enterprise information systems) have been developed. In order to ensure high quality of the delivered products, the accumulated knowledge has been consolidated and transformed into company-wide development standards and a specific architectural style termed Quasar (Quality Software Architecture [16]). It has been extended over the years into a method that covers all disciplines of software development projects. In its current version, Quasar comprises integrated methods for business modeling, requirements engineering, analysis, design, implementation, deployment and test of individual IT applications. During the last years and as an answer to increasing customer demands for enterprise-wide service-oriented application landscapes, new concepts and methods have been developed. They are clustered in a new architectural style and development method termed Quasar Enterprise [4, 5].

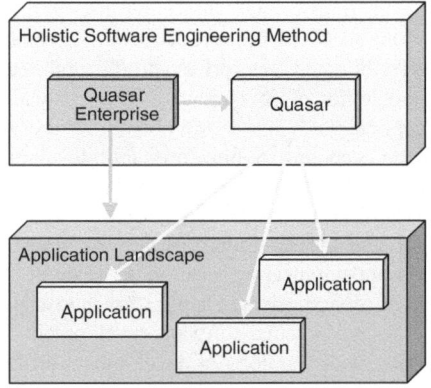

Fig. 1. The proposed holistic software engineering method integrates the application landscape-oriented development method Quasar Enterprise with the application-oriented development method Quasar

An important distinction between Quasar and Quasar Enterprise is their focus of development: While Quasar focuses on individual applications, Quasar Enterprise focuses on whole application landscapes. In order to realize an overall development of application landscapes including the development of their contained applications, both development methods have to be integrated. Only such a holistic view on the development will enable a seamless development process of large application landscapes (see Fig. 1). Thus, building service-oriented enterprise architectures demands for new software development methodologies, too. They need to integrate business modeling, application landscaping and application software development. Software companies are thus undergoing an enterprise transformation as well.

This integration of methods is not only important for continuous development of application landscapes, it also acknowledges the need for gentle migration from existing enterprise applications towards service-oriented enterprise application landscapes. This is necessary since transforming enterprises to follow a service-oriented paradigm and accordingly changing their business and IT landscapes into SOA needs time and diligence. On the path towards the ultimate service-oriented enterprise architecture, hybrid (i.e., partly non-service-oriented and service-oriented) enterprises evolve for the transition period. The same observation holds for software companies: changing of development methods in a large software company can only be done with caution, especially for a paradigm shift such as the move from development of individual applications to the development of service-oriented application landscapes.

In this paper, we describe in particular from a technical perspective how the application-oriented development method Quasar is integrated with the application-landscape oriented method Quasar Enterprise in order to yield a holistic, service-oriented software engineering method. We illustrate how a company-wide ontology of development artifacts serves as the key feature for integrating both methods. The integrated method of application-landscape development becomes the driving force of an enterprise transformation in the software company itself. Particularly, the strengthening of the business modeling discipline as well as the tight integration of application landscaping and software development cause a change in the enterprises that produce the software.

The paper is structured as follows: In Sect. 2, we start with a brief characterization of our view on service-oriented application landscapes, the target of development of the presented holistic software engineering method. This is done due to the observation that quite diverse understandings of service-orientation exist in the literature as well as in practice. Sections 3 and 4 summarize the existing development methods Quasar Enterprise for application landscapes as well as Quasar for single applications. Section 5 illustrates the integration approach for yielding a holistic method. It is shown how the underlying integrated Quasar Ontology serves as the integration base. Section 6 summarizes the achievements and discusses future work.

2 Evolution towards a Service-Oriented World

In our modern and fast changing world, enterprises must have a high degree of flexibility in order to react to changing demands of markets and customers. They must be able to quickly adapt their business to the changing world to be competitive. This means that their business processes need to be flexible and adaptable. In turn, this can only be put into practice if the supporting IT systems have the same flexibility: If business processes change, the IT systems must be changed accordingly. It must be possible to change or reassemble existing processes, add new services, or alter existing services without too much effort and in short time. Enterprises are thus enabled to change their business e.g. by adding new products or services to their portfolio, optimizing their business and production processes, re-organizing the enterprise's structure and responsibilities of organizational units, or changing the enterprise's in-house production depth by outsourcing and insourcing, offshoring and reshoring. The business changes must be reflected in equivalent adaptations of the supporting IT systems without major restructuring of the enterprise IT architecture (see [4,5]).

Figure 2 illustrates this view on a service-oriented architecture. Business processes (at the top of the figure) are supported and realized by services, which bridge the process layer with the underlying layer of component-based IT applications.

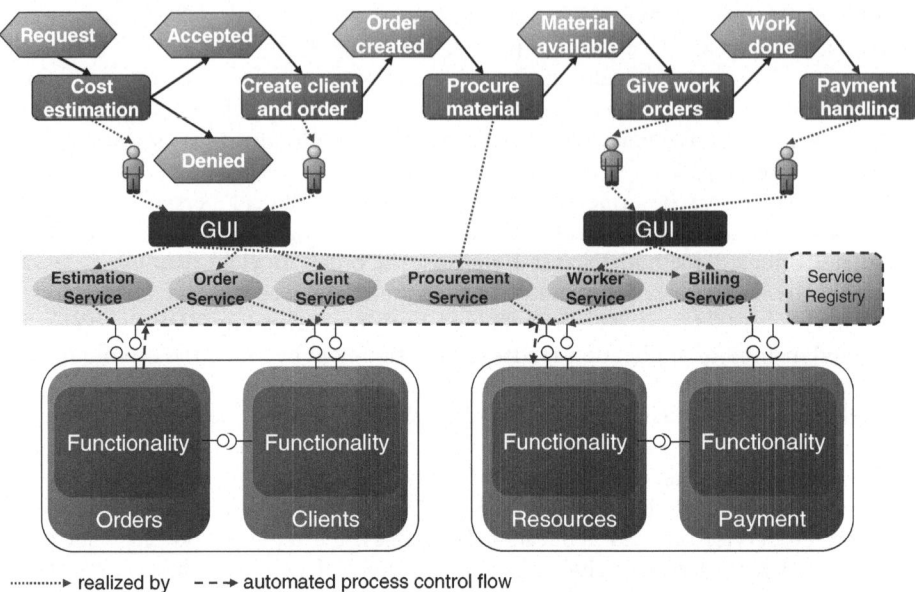

.............► realized by   – – → automated process control flow

Fig. 2. Service-oriented application landscape (from [3])

Service-oriented architectures are a promising approach for these requirements. They provide a paradigm for structuring both the business of an enterprise and the enterprise IT architecture. The business is structured according to services, and the IT architecture is structured according to these business services.

Yet, only a holistic view on the development and maintaining of service-oriented application landscapes including their applications provides such flexibility and adaptability in case of changing requirements or context constraints. Therefore, development methods have to be transformed and integrated as well in the context of a service-oriented paradigm.

The upcoming two sections will summarize the existing methods Quasar Enterprise and Quasar, before we explain the integration approach in Sect. 5.

3 Quasar Enterprise

The Quasar Enterprise (QE) method [4, 5, 8] provides detailed guidelines for developing service-oriented enterprise architectures. Other approaches for developing service-oriented architectures are described e.g. in [11] by Krafzig, [7] by Erl, [2] by Dostal, and [18] by Woods.

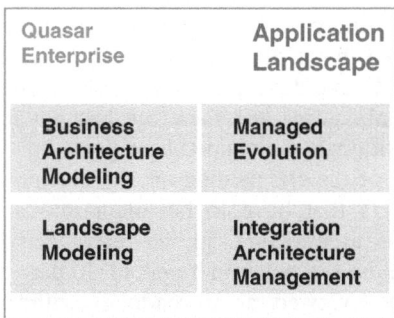

Fig. 3. Disciplines of the Quasar Enterprise Method

Quasar Enterprise consists of four disciplines (see Fig. 3):

1. Business Architecture Modeling
2. Landscape Modeling
3. Managed Evolution
4. Integration Architecture Management

In particular, the first two disciplines realize a top-down approach for identifying and designing services.

In Business Architecture Modeling, the business architecture is analyzed and defined. The QE method suggests starting with business goals and deriving architectural guidelines from them. Afterwards, business services are identified and described.

In Landscape Modeling, the ideal application landscape is defined. It is termed 'ideal' as it does not take into account constraints like existing application components, but only aims at optimal fulfillment of business goals by appropriate services. The first step in this discipline is the definition of domains as the underlying business-driven structure of IT applications [8]. Application services are derived from the high-level business services, are assigned to domains and grouped in suitable (logical) application landscape components. The application landscape components' interfaces and operations are identified and the communication between components is described as a coupling architecture.

The discipline Managed Evolution of the QE method provides guidelines how to structure the evolution process from an existing IT landscape into the direction of the identified ideal landscape. Based on the evaluation of the as-is architecture and the ideal architecture, the target architecture for each evolution step is determined.

Here, the so far top-down approach of QE is coupled with a bottom-up approach which looks for existing applications and determines their appropriateness for realizing identified application landscape components. In particular, it is decided whether existing applications are still usable as realizations of identified application landscape components or whether new applications have to be developed.

We will explain below that this is the first reference point in the envisaged holistic software engineering method where the application landscape development method Quasar Enterprise is integrated with the development method Quasar for individual applications.

The final discipline Integration Architecture Management of QE takes care of the integration architecture and the right choice of a concrete integration platform. This part of the application landscape development method is independent of the choice of developing or reusing applications and therefore has not to be changed within the holistic method. So it is not further explained here.

This rough description of the QE method and its disciplines gives an overview of the most important aspects that have to be taken into account when planning a service-oriented enterprise architecture. In the referenced literature (ref. [4, 5, 8]), Quasar Enterprise is explained in much more detail. In particular, it is illustrated how Quasar Enterprise can be viewed as a roadmap within the generic Integrated Architecture Framework (IAF) [9]. The next section will introduce the Quasar method for application development.

4 Quasar

Quasar (Quality Software Architecture [16]) has been developed as the standard architecture of Capgemini sd&m. Originally, Quasar was focused on software architecture and how to conceptualize, design and build information systems based on this architecture. In recent years, Quasar has been further extended in scope to become a software engineering method that covers all disciplines of software development projects. Quasar is now the central part of Capgemini sd&m's method portfolio (see Fig. 4). Other methods have evolved around Quasar such as the project management method ePM (efficient and effective project management). Additionally, Quasar Infrastructure provides the dedicated tool support for all methods in the form of an integrated tool suite. All methods together cover the full scope of disciplines of the Rational Unified Process (RUP) [12].

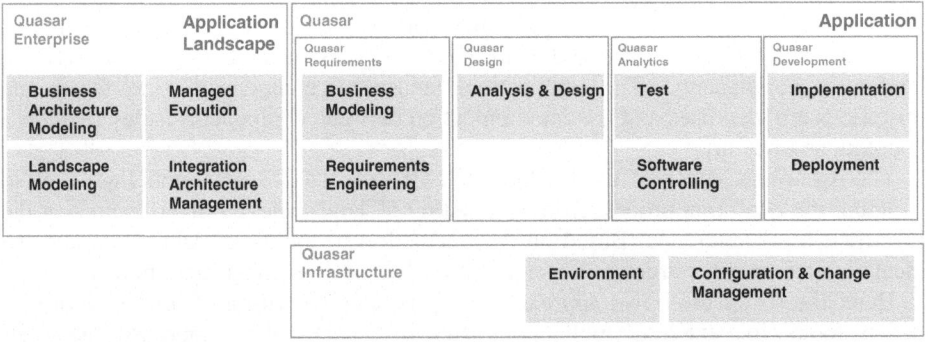

Fig. 4. The disciplines of Quasar and Quasar Infrastructure

Alike RUP disciplines, the methods are logical containers that group the process elements such as roles, activities, artifacts, and workflows, the associated concepts, guidelines, templates, roadmaps and advice on tool usage, and provide process guidance. They have been developed in consideration of established standards such as RUP, The Open Group Architecture Framework (TOGAF) [17] and the Integrated

Architecture Framework (IAF) [9], the International Software Testing Qualifications Board (ISTQB) standards [10], and the Project Management Body of Knowledge of the Project Management Institute (PMI) [13].

Quasar covers the six engineering disciplines of RUP together with a seventh one, Software Controlling. It is further divided into four parts: requirements, design, development, and analytics:

- Quasar Requirements contains the two disciplines Business Modeling and Requirements Engineering. The former is concerned with the production of the business vision and the business model restricted to the scope of an individual IT application. The latter deals with the system vision and the analysis, elicitation, documentation, validation and management of application requirements (functional requirements and quality requirements), organizational requirements, and integration requirements.
- Quasar Design contains the discipline Analysis and Design. This discipline comprises system specification (i.e., system analysis and conceptual design), technical design, system architecture (including architecture views and principles), design and implementation patterns.
- Quasar Development contains the two disciplines Implementation and Deployment. Coding principles, system reengineering, and software and component frameworks are the subject of the Implementation discipline. Deployment is concerned with the construction of pilot versions, releases and the rollout process as well as with issues of installation and migration.
- Quasar Analytics is the analytical counterpart to the three aforementioned constructive parts. It contains the two disciplines Test and Software Controlling. The former covers the whole process of software and system testing across unit, integration, system, and acceptance testing stages. The Test discipline considers the test method, test management, test frameworks, and issues of acceptance (e.g. acceptance criteria with respect to requirements). Software Controlling is concerned with the continuous quality monitoring and control throughout the development process. It defines a quality assurance process through audits (so-called Quality Gates (see [14, 15]) and the Software Cockpit [1] for continuously monitoring the progress of software and accompanying artifacts.

Quasar Infrastructure is concerned with both the environment as well as configuration and change management. They constitute the other two supporting disciplines in RUP.

In order to end up with a holistic software engineering method, which covers both application landscaping and application development methods, Quasar and Quasar Enterprise need to be integrated. The common ground for integrating all methods is given by the Quasar Ontology which will be explained in the next section.

5 Integrating Quasar Enterprise with Quasar

After having briefly sketched the two existing methods Quasar Enterprise and Quasar, we will now explain our approach for integrating them into a holistic software engineering method. The overall idea is to base both methods on a common underlying ontology, which will not only ensure a consistent understanding of notions of Quasar Enterprise and Quasar, but will also serve to enable the transition between the application landscape view of Quasar Enterprise and the individual application view of Quasar. By reference to structural as well as behavioral aspects of an application landscape and an application, we show their integration. We will provide examples for both aspects.

5.1 Quasar Ontology

Defining the ontology of development artifacts is not only necessary for the integration of Quasar Enterprise and Quasar, but essential for mastering the challenges of industrial software development. Industrial software development has to deal with world-wide distributed development where the development processes are artifact centered. Artifacts are normally the mostly used communication channel in distributed development. Therefore, the quality of all development artifacts has to be high and it is necessary to have highly efficient development processes.

The team and project situation is often in conflict with these characteristics:

- different culture within business units or branches, especially in right-shore projects,
- diverse understanding of notions, concepts, terminology, languages,
- diverse structure, content, and maturity of development artifacts,
- diverse understanding of purpose of development artifacts,
- diverse development processes,
- diverse tool support, and
- diverse background of team members.

Therefore, it is important to precisely define notions and their interrelations, to cluster notions into artifacts according to the software development disciplines, and to present the notions and their interrelations in a uniform and comprehensible way (see Fig. 5). This generic structure of a software development method has been developed as contribution in the area of method engineering [6]. It will be used here to integrate the two existing methods Quasar Enterprise and Quasar.

The Capgemini sd&m common reference model of notions is called Quasar Ontology. It enables common understanding of notions and developed artifacts even between cultures, supports seamless transition between the disciplines within the development process, allows for standardized processes based on the reference model, and establishes the meta-model for providing tool support.

Figure 5 shows the elements of the generic software engineering method [6]. Elements of Quasar Enterprise are shown in yellow (light gray in print), elements of Quasar are green (dark gray in print), integrated elements are two colored. More precisely, the development process is described by defined development methods and their building blocks (modules), including support by precise guidelines, patterns, reference architectures and scenarios. The focus of the development methods is on the elaboration of artifact types. The artifact types cluster notions and appropriate languages. The used notions and their interrelations are given within the ontology. Tool support is available for the method elements. How our holistic software engineering method is built on this ontology is explained in the following sections.

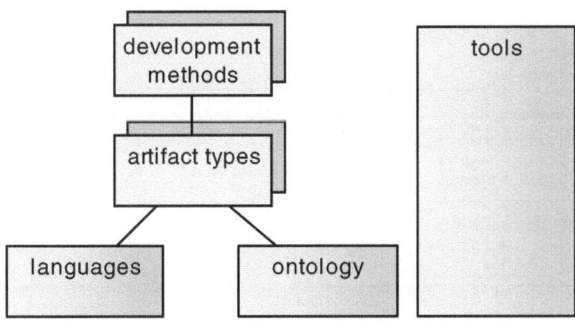

Fig. 5. Elements of a generic software engineering method

5.2 Structure

First, we focus on the structural aspects of application landscapes and applications. Both Quasar Enterprise and Quasar define artifacts for structuring an application landscape and application, respectively.

Within Quasar Enterprise the following notions are defined: Application Landscape, Domain, and Logical Application Landscape (short: AL) Component; whereas Quasar uses the notions: Application, Subsystem and Logical Application Component.

An Application Landscape denotes the entirety of application systems, which an enterprise operates for organizing and completing its business. Normally, the application systems do not stand for themselves alone, but are networked over common data bases or interfaces. In the Quasar Ontology, the Application Landscape consists of Domains in the Quasar Enterprise view and Applications in the Quasar view. Domains are used to hierarchically group Logical AL Components according to business aspects. Applications bundle Subsystems and Subsystems hierarchically bundle Logical Application Components with respect to functional aspects – to keep the application's complexity under control. Subsystems are normally finer grained than Domains.

In Fig. 6, the integration of Quasar Enterprise and Quasar is defined by interrelating the corresponding parts of the ontologies of these two methods. The left-hand side shows Quasar Enterprise's artifacts that are used to structure an enterprise's business and IT architecture, whereas the right-hand side contains Quasar's artifacts that are used to structure applications. The relationships between Quasar Enterprise's and Quasar's artifacts are defined as follows: First, an Application Landscape may consist of Domains (on the QE side) as well as of Applications (on the Quasar side).

The more interesting relationship is between Logical AL Components and Quasar Structure Elements. Quasar Enterprise offers a method for finding Logical AL Components. This is done by decomposing Business Services down to Elementary Business Services (see Fig. 8). In this process, the Domains are identified. Thereafter, the Elementary Business Services that will be automated by an Application Function (on the Quasar side) are selected and categorized. Application Functions of the same category are thus bundled in a Logical AL Component. For a detailed description we refer to [5].

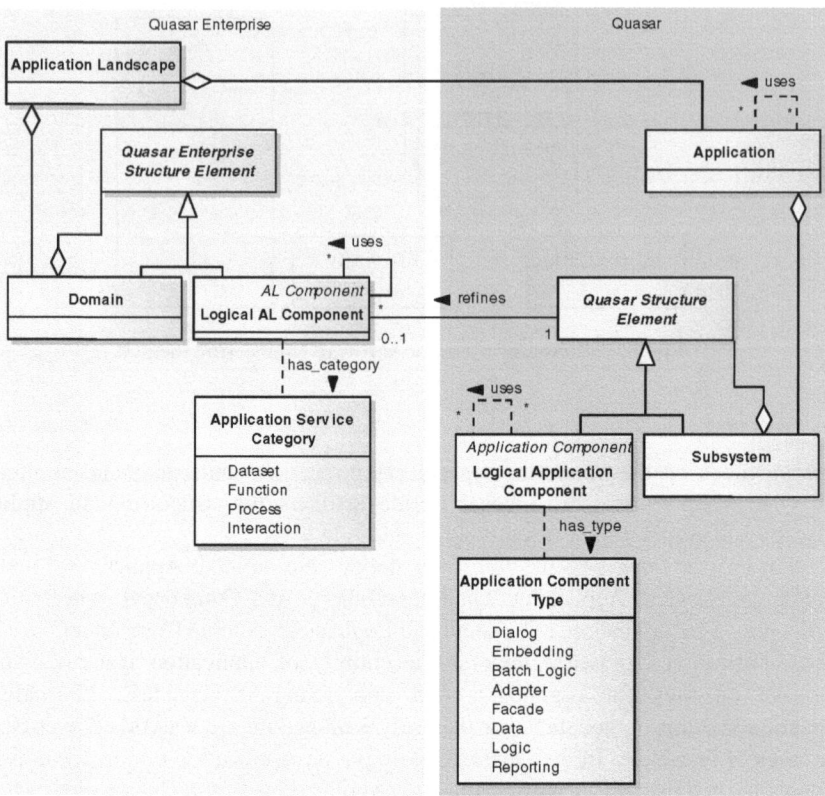

Fig. 6. Excerpt from the integrated ontologies of Quasar Enterprise and Quasar: structural part

Within step 3, i.e., the Managed Evolution step of the Quasar Enterprise method, it has to be decided whether such a Logical AL Component is realized by reusing an existing application or by developing a new one. This is expressed in the integrated ontology (see Fig. 6) by the 'refines' edge between Logical AL Component on the Quasar Enterprise side and the Quasar Structure Element (with a Subsystem at the root) on the Quasar side. If the application has to be newly developed, the Quasar method will be used for its development. This means that the Logical AL Component is regarded as a subsystem and structured and refined possibly into subsystems and finally into Logical Application Components.

The segmentation into Subsystems can be process-oriented, organization-oriented or data-oriented. In process-oriented segmentation, the Subsystems are based on functional relationships in the business area supported by the application. Organization-oriented segmentation is based on the entities in the customer's organization structure. In case of data-oriented segmentation, the assignment is based on the coherence of entity types in the logical data model. Subsystems are created according to highly cohesive entity types in the logical data model and the Application Functions that operate on the same. This means Subsystems created in this way are based on data sovereignty over the entity and relationship types of the logical data model.

Concluding, this 'refines' edge between the Quasar Enterprise and the Quasar part of the integrating Quasar Ontology does not only relate notions, but also links methodical building blocks of Quasar Enterprise and Quasar. Altogether, the integration and interrelation of notions and methodical building blocks yield the holistic and integrated software engineering method.

5.3 Example: Structure

In this section, we give an example how an application landscape can be structured and how a smooth transition between global application landscaping and local application development is achieved. For reasons of nondisclosure we had to alter some information. The example is taken from an enterprise in the accounting sector.

In Fig. 7, the structure of the enterprise is shown. We use the UML package symbol for depicting Domains and Subsystems, whereas the UML component symbol is used for Logical AL Components and Logical Application Components. First, the application landscape is split according to the main Business Services. Amongst others, resulting Domains are the customer communication, services, sales and accounting domains. Looking at the accounting domain, analyzing the sub-ordinate Business Services and Business Objects led to identifying some Logical AL Components. One Logical AL Component is the 'Manual Adaption and Approval Process' component. Its category is interaction, as the contained application services are interactions with the user. The component provides the user access to other components in the accounting domain.

Relating this Logical AL Component from the Quasar Enterprise view to a Subsystem in the Quasar view, it can be refined and structured further. The structuring is now done with the help of Quasar methodical steps on use-case level and no longer with Quasar Enterprise methodical steps on business-service level. Finally, this ends up in the identification of Logical Application Components. For example, within the subsystem 'Manual Adaptation' one sees two components. One is the 'Manual Adaptation Dialog' component of type 'Dialog'. This component is a client-side component and contains the dialog functionality. In contrast, the 'Adapt Accounting Data' component is of type 'Logic'. This component is a server-side component and contains the business logic for adapting accounting information.

5.4 Behavior

In this section, we concentrate on the description of behavior in Quasar Enterprise and Quasar. Quasar Enterprise uses Application Services to describe the external behavior of an application system, whereas Quasar uses Use Cases. The questions we asked for setting up the holistic software engineering method were: Are Application Services and Use Cases connected? Can we use the same methods and languages for describing Application Services as we use for Use Cases? Can we use the same artifacts (e.g. dialogs or application functions) for detailing Application Services as we use for Use Cases? In this section we show that it is possible to use the concepts of Quasar to detail the artifacts of Quasar Enterprise.

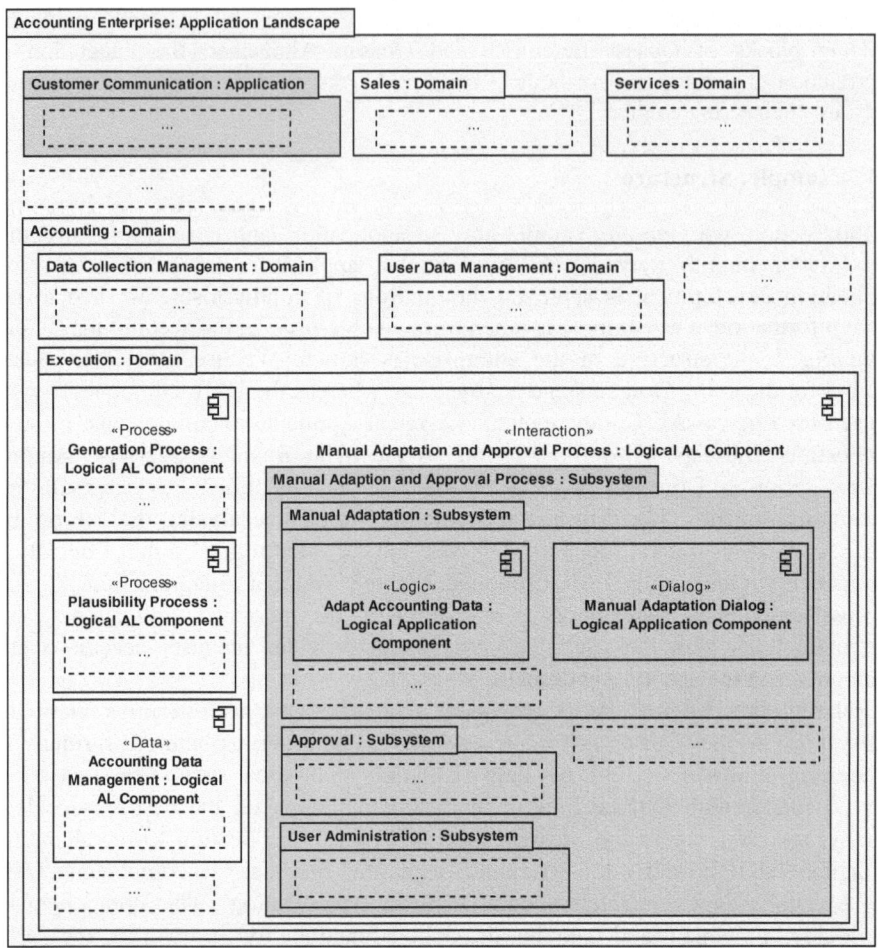

Fig. 7. Example of relating structural aspects

In Fig. 8, Quasar Enterprise's artifacts are on the left-hand side and Quasar's artifacts are on the right-hand side. One of Quasar Enterprise's methods is decomposing Business Services down to Elementary Business Services which are realized by Logical AL Components. One of Quasar methods is decomposing Business Processes down to Use Cases which are realized by Logical Application Components. Comparing these two sides and interrelating the notions of both sides enables here, too, a smooth transition from global application landscaping down to local application development. The interrelations between both sides are described by appropriate 'refines' relationships. In summary, Business Service Activities can be viewed as the external view of an Activity (i.e., Business Process or Task).

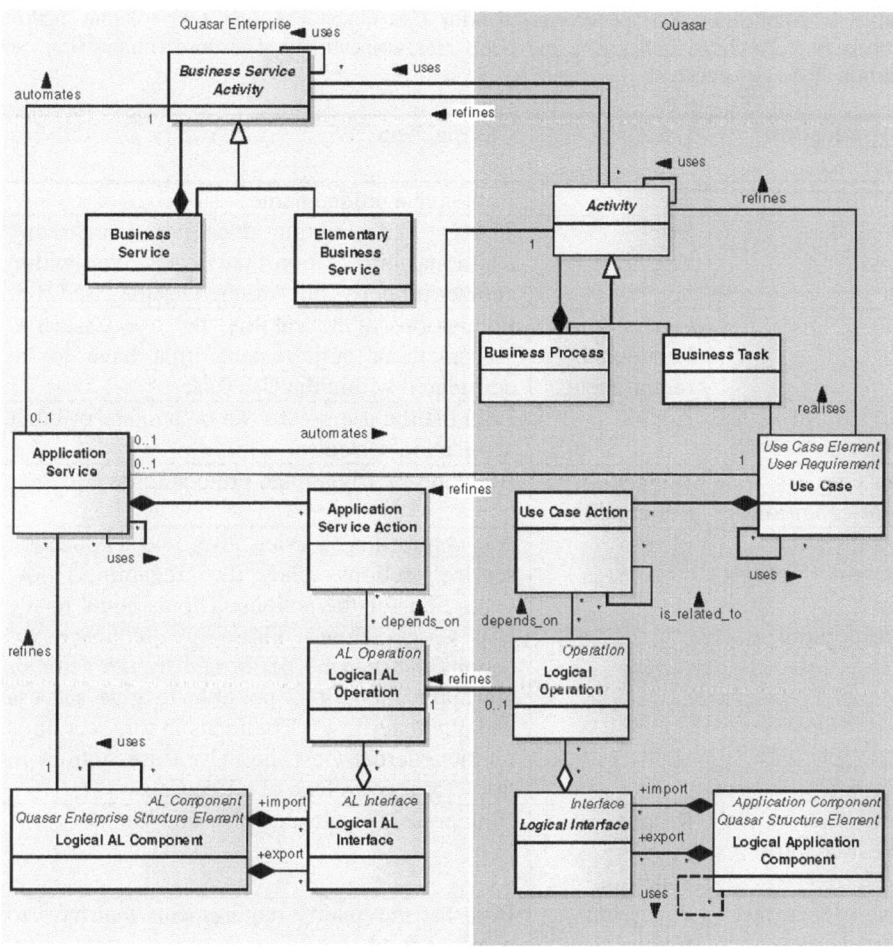

Fig. 8. Excerpt from the integrated ontologies of Quasar Enterprise and Quasar: behavioral part

Investigating these interrelationships in more detail, one can state the following: Quasar Enterprise uses Application Services to define the Business Service Activities that are automated by an application system, whereas Quasar takes Use Cases to describe Business Tasks that are supported by an application system.

Table 1 depicts the properties of Application Services versus Use Cases, and shows that Application Services can be described with the same concepts as used for Use Cases. Furthermore, it is possible to use Quasar's artifacts and methods to specify, design and implement Application Services. Thus, a comparison and integration of the two methods Quasar Enterprise and Quasar based on a common ontology does not only facilitate the smooth transition between global landscaping and local application development, but also supports the reuse of well-experienced techniques in the application development in the development of service-oriented application landscapes, contributing to a gentle transformation within the software company.

Table 1. Application Service contrasted with Use Case. The first two columns list the properties of Application Services and Use Cases, respectively. The third column states the relation of the two concepts.

Application Service	Use Case	Comparison
Name	Title	Both give a unique name.
	Brief description	The brief description does not contain any additional information that is not given within another property, but summarizes the Use Case.
	Objective / Functional requirements	For reasons of traceability, the Use Case lists all functional requirements that have to be considered within the Use Case.
User of the Service	Actors	Both list the users who can be humans or other application systems.
Trigger / Pre-condition	Trigger / Pre-condition	Both present trigger and pre-condition.
Application service actions and flow description	Flows	An application service lists the application service actions and the regarding flow restriction for the actions. This is equal to the Use Case's flows. Flows are sequences of actions that can be performed by the actor or the application. It is possible to give success and alternate flows. The focus in this section is on the external view, i.e., Use Case Actions in the actor's swim lane of a Use Case.
Results / Post-condition	Results / Post-condition	Both present results and post-conditions.
Non-functional requirements	Execution frequency / Quality requirements	Both list the quality requirements that have to be considered.
Process description	Flows	A process description is again equal to the Use Case's flow. The focus in this section is on the internal view, i.e., Use Case Actions in the application's swim lane of a Use Cases.

5.5   Example: Behavior

In this section, we come back to the accounting enterprise example. In Fig. 9, we show the connections between concrete Quasar Enterprise and Quasar artifacts.

The Business Service represents the external view and the Business Process the internal view of 'Manual Adaptation and Approval' functionality. As the figure shows, not all Use Cases have to be defined as Application Services, but it is possible to take Use Cases for refining Application Services. Once we have transformed the functionality in a use-case view, we can use the established Quasar methods further on to implement it.

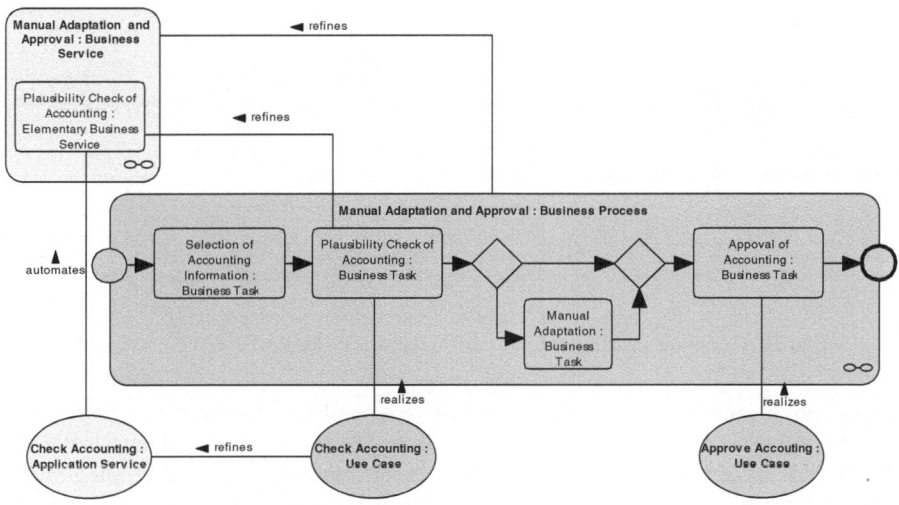

Fig. 9. Example of relating behavioral aspects

6 Conclusions

In this paper, we have outlined that the development and management of an IT application landscape as part of a service-oriented enterprise architecture requires a holistic approach, which reaches from business modeling and global application landscaping down to the local development of individual software applications. We have illustrated how such a holistic software engineering method can be systematically composed and integrated from existing methods for developing application landscapes and individual applications, respectively.

Based on previous research results in the area of method engineering [6], we have shown how two concrete existing methods can be integrated. In particular, we used a common ontology where refining links interrelate relevant notions of both methods. By this, a smooth transition between both methods could be defined.

The approach has been exemplified with two existing methods, which are developed and used by the software company Capgemini sd&m. These are Quasar Enterprise for developing application landscapes and Quasar for developing single applications (see Fig. 10).

Overall, we have shown that a transition of enterprises towards service-oriented enterprises impacts the way how enterprise architectures and their IT support are developed and managed. Thus, this has direct impact on software companies and their development methods leading to a transition of these enterprises, too.

While the here presented integrated and holistic software development method has already been successfully applied in several software development projects, the method might be further improved. In particular, it has to be investigated how applications (components) can be reused and integrated in case of limited knowledge of the application's (component's) functionality. This holds true in the case where packaged solutions (like SAP modules) have to be included in an application

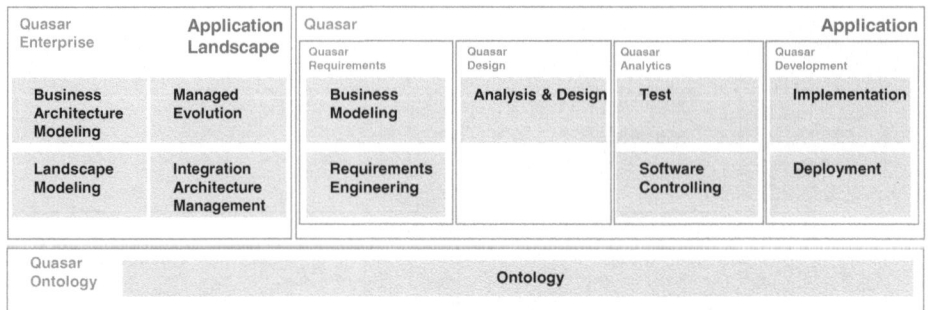

Fig. 10. Common Quasar Ontology for integrating Quasar Enterprise and Quasar

landscape. Furthermore, in the presence of federated development processes, several development methods might have to be integrated, as individual applications are developed according to different development methods. Last, the integration of different methods and their development artifacts requires consistency management of these artifacts, e.g. in the case of dynamic change of business processes.

References

1. Bennicke, M., Steinbrückner, F., Radicke, M., Richter, J.-P.: Das sd&m Software-Cockpit: Architektur und Erfahrungen. In: Koschke, R., Herzog, O., Rödiger, K.-H., Ronthaler, M. (eds.) INFORMATIK 2007. Proc. 37. Jahrestagung der Gesellschaft für Informatik e.V (GI), vol. 2, LNI, vol. 110, pp. 254–260. GI, Bonn (2007)
2. Dostal, W.: Service-orientierte Architekturen mit Web Services: Konzepte - Standards - Praxis, 1st edn. Elsevier Spektrum Akad. Verl, Munich (2005)
3. Engels, G., Assmann, M.: Service-Oriented Enterprise Architectures: Evolution of Concepts and Methods. In: Proc. 12th International IEEE Enterprise Distributed Object Computing Conference (ECOC 2008), pp. xxxiv–xliii. IEEE Computer Society Press, Los Alamitos (2008)
4. Engels, G., Hess, A., Humm, B., Juwig, O., Lohmann, M., Richter, J.-P., Vo , M., Willkomm, J.: Quasar Enterprise: Anwendungslandschaften serviceorientiert gestalten. dpunkt, Heidelberg (2008)
5. Engels, G., Hess, A., Humm, B., Juwig, O., Lohmann, M., Richter, J.-P., Vo , M., Willkomm, J.: A Method for Engineering a True Service-Oriented Architecture. In: Cordeiro, J., Filipe, J. (eds.) Proc. 10th Intl. Conf. on Enterprise Information Systems (ICEIS 2008), vol. ISAS-2, Barcelona, Spain, pp. 272–281 (2007)
6. Engels, G., Sauer, S., Soltenborn, C.: Unternehmensweit verstehen – unternehmensweit entwickeln: Von der Modellierungssprache zur Softwareentwicklungsmethode. Informatik-Spektrum 31(5), 451–459 (2008)
7. Erl, T.: Service-oriented Architecture: Concepts, Technology, and Design. Prentice-Hall, Upper Saddle River (2006)
8. Hess, A., Humm, B., Vo , M., Engels, G.: Structuring Software Cities - A Multidimensional Approach. In: Proc. 11th IEEE Intl. Enterprise Distributed Object Computing Conference (EDOC 2007), pp. 122–129. IEEE Computer Society Press, Los Alamitos (2007)

9. Integrated Architecture Framework (IAF), Capgemini,
   `http://www.capgemini.com/iaf`
10. International Software Testing Qualifications Board (ISTQB),
   `http://www.istqb.org/`
11. Krafzig, D., Banke, K., Slama, D.: Enterprise SOA: Service-oriented Architecture Best Practices. Prentice-Hall, Upper Saddle River (2006)
12. Kroll, P., Kruchten, P.: The Rational Unified Process Made Easy: A Practitioner's Guide to the RUP. Object Technology Series. Addison-Wesley, Reading (2003)
13. PMBOK Guide: A Guide to the Project Management Body of Knowledge, 4th edn., Project Management Institute (2008), `http://www.pmi.org/`
14. Salger, F., Bennike, M., Engels, G., Lewerentz, C.: Comprehensive Architecture Evaluation and Management in Large Software-Systems. In: Becker, S., Plasil, F., Reussner, R. (eds.) QoSA 2008. LNCS, vol. 5281, pp. 205–219. Springer, Heidelberg (2008)
15. Salger, F., Sauer, S., Engels, G.: Integrated Specification and Quality Assurance for Large Business Information Systems. In: Proc. 2nd India Software Engineering Conference (ISEC 2009), Pune, India, February 23-26 (2009) (to appear)
16. Siedersleben, J.: Moderne Softwarearchitektur - Umsichtig planen, robust bauen mit Quasar. dpunkt, Heidelberg (2004)
17. The Open Group Architecture Framework, TOGAF 8.1.1 'The Book, The Open Group (2007), `http://www.opengroup.org/architecture/`
18. Woods, D., Mattern, T.: Enterprise SOA: Designing IT for Business Innovation. O'Reilly, Beijing (2006)

Empowering Full Scale Straight Through Processing with BPM

Eric D. Schabell[1] and Stijn Hoppenbrouwers[2]

[1] SNS Bank, Postbus 70053, 5201 DZ 's-Hertogenbosch The Netherlands
[2] Institute for Computing and Information Sciences, Radboud University Nijmegen
Toernooiveld 1, 6525 ED Nijmegen, The Netherlands

Abstract. The SNS Bank (the Netherlands) has made a strategic decision to empower her customers on-line by fully automating her business processes. The ability to automate these service channels is achieved by applying Business Process Management (BPM) techniques to existing selling channels. Both the publicly available and internal processes are being revamped into full scale Straight Through Processing (STP) services. This extreme use of online STP is the trigger in a shift that is of crucial importance to cost effective banking in an ever turbulent and changing financial world. The key elements used in implementing these goals continue to be (Free) Open Source Software (FOSS), Service-oriented architecture (SOA), and BPM. In this paper we will present an industrial application describing the efforts of the SNS Bank to make the change from traditional banking services to a full scale STP and BPM driven bank.

1 Introduction

The SNS Bank in the Netherlands is making a strategic move to automate her support and selling channels to provide her customers with modern on-line services. Realizing that it will take more than just an on-line web shop to excel in the financial world, she has also moved to automate many internal processes. The key elements used in implementing these goals are full scale Straight Through Processing (STP) [1] and Business Process Management (BPM) [2].

In this paper we will present the efforts made to change from traditional banking services to a full scale STP and BPM driven financial institution. We begin in section 2 by clarifying what *full scale STP with BPM* means and why this is of importance to the future of SNS Bank. In section 3 we take a closer look at our case study, the *STP Purchasing* project. We will provide some insights into the application of STP with BPM within an open source development environment, discuss the component architecture, take a look at our process modelling steps, examine how we utilized customer testing, and conclude with an overview of some general empirical data. We will present our experiences, both good and bad, in dealing with a large BPM implementation. As can be expected, there will always be challenges to be met when such an expansive shift in strategy is being implemented and we start our tour in section 4 of the issues encountered

E. Proper, F. Harmsen, and J.L.G. Dietz (Eds.): PRET 2009, LNBIP 28, pp. 18–33, 2009.

in the project. Section 5 will discuss the brighter side, outlining the positive impact that this project has had on both business and technical realms. This will leave the reader with a good idea of the challenges involved, hopefully helping in implementing other industry BPM applications. We will take a into the futrue in section 6 and provide an overview of ongoing development. Section 7 is a look at moving ahead while applying the lessons we have learned.

2   Full Scale STP

The application of STP with BPM is not a new phenomenon in the financial industry, with other banks having reported some success with relatively straight forward on-line financial solutions [3,4]. Some are even dreaming of taking on the more challenging processes within the banking industry, such as mortgage processes [5]. The difference between these types of solutions and the one presented here concerns complexity. We offer the follwoing definitions:

Definition 1 (Business Process Mangement). *Business Process Mangement is focused on aligning business processes to the customers want and needs by applying methods, tools and solutions.*

This is a simple and straight forward look at how we intend to apply BPM within our organization.

Definition 2 (Straight Through Processing). *Processing a business transaction automatically, without requiring people to be involved in the process. The purpose of STP is to create efficiencies, eliminate mistakes, and reduce costs by having machines instead of people process business transactions.*

This definition is in line with most of the definitions we have encountered in the financial world [6,7,8]. It will work fine as a beginning definition of how we construct our processes, but we need to refine it a bit for real world financial business processing:

Definition 3 (Full Scale STP). *A straight through process (STP) implementation that requires the solution to encompass a wide range of system integration and will include human tasks which embody the complex decision making that automation either cannot legally implement, or is precluded by technical limitations.*

We exclude cost as a factor to determining if an implementation is full scale STP or not. We feel that cost, in terms of time, money, or other value risk, is a business concern that is not related to complexity, but rather to some current operational or environmental situation (i.e. budgets, deadline pressures, politics, environment, etc.).

The drive to push for full scale STP with BPM is multifaceted. The leading goals are cost reduction, manpower reduction in business processes, removing potential (human) mistakes, and channel independent processing. Users should experience such processes as transparent, quick, simple, directly usable, and should be able to complete their task in one attempt.

SNS Bank is targeting effective and efficient processing where as much human intervention as possible has been removed. The customer will be kept informed at crucial process steps, communication always being an important factor in customer experience. For the cases that are exceptions or fall out of STP processing, there will be clear and predefined processes to ensure expeditious handling. Last but not least, the entire communication process is as paperless as can be. This encapsulates the SNS Bank's idea of full scale STP processing.

As Heckl and Moormaan [9] conclude "...long term success cannot be achieved without the development of new business ideas, innovative products and services, and customer retention." We believe that such success can only be achieved if BPM techniques are fully integrated. Full scale STP with BPM will continue to be expanded on and implemented throughout the range of products, sales channels, and business processes that affect both customer and customer support. We believe that the time for full scale STP with BPM is now.

3   A Case Study

In the beginning of 2007 the first full scale STP project at SNS was launched, with the goal of putting four new savings products on-line at the start of 2008. This project is known as *STP Purchasing* and will provide us with a case for closer examination of full scale STP with BPM. This section will present the component architecture, take a look at how the process was modelled, show how customer testing was used to verify the solution, and provide some empirical data on the results.

3.1   Overview

The goals for this project were: for a customer to be asked as few questions as possible during the purchasing process; that the entire process would be completed within a maximum of five clicks in the on-line website; and that the customer would be kept informed during all crucial steps in the process with clear, directed communication relevant to a specific purchasing process. A further desire was to maximize paperless communication with the customer. It was essential to maintain as short a processing time as possible, with processes that land in human action stages causing no more than one day delay. It should be volume independent, deliver reusable processes, reusable services, be multi-label, and multi-channel. Above all, the project should provide a full scale STP solution with a maximum degree of automation.

With our definition of full scale STP [definition 3] in mind, we already have an idea that the process is not free from human tasks. There are several instances in which we could not avoid having human interaction as part of this process. The resulting challenges will be discussed in more detail later on in this section.

The project resulted in a general end-to-end purchasing process, initially for savings products, and a new process for document scanning and storage.

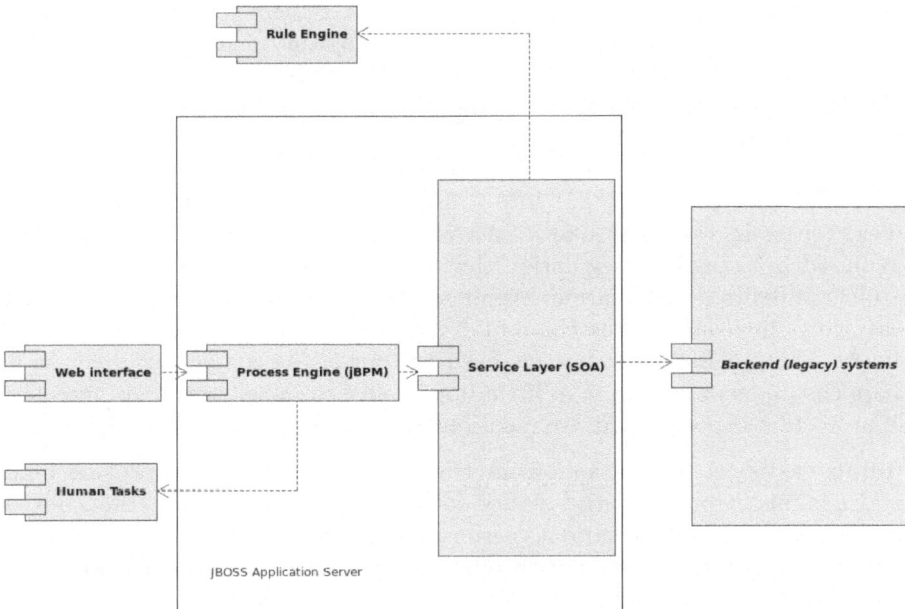

Fig. 1. STP Purchasing architecture

A purchasing request database implementing the data model for each processing request was delivered along with a BPM process flow; a web front end was
created for the initial savings products, and the relevant SOA services. A new
department was created, called *Process Management Evaluation and Processing*.
Total project IT investment was 14,000 hours.

3.2   Architecture

The SNS implementation environment for full scale STP with BPM is one of
pure Java [10]. The emphasis is on building solutions within the bank's own IT
department, making use of Free Open Source Software (FOSS) where possible,
achieving reusability of existing applied solution components, and using best of
breed components when forced to shop outside of our existing code base.

There was a shift in component strategy in 2004 from three main commercial suppliers to one where FOSS components are preferred when possible. Open
source is now quite pervasive throughout the solution architecture of all SNS
projects. Furthermore, the development environment and tooling used to implement the solution consists of almost only FOSS. This is outside the scope of this
paper and will therefore be excluded from further discussion. The component
architecture as shown in Figure 1 is a very generic and high level view. We will
discuss the components as shown, from left to right.

Web interface. The entry point for any full scale STP application is the web
interface as seen by the customer in the on-line banking website. This is a Java

based website that makes use of a content management system. In the STP Purchasing project it provides the user with the option to apply for one of four saving products. If placed, a request is gathered together with user information, verified through various web services, and then using a web service it is deposited into the *Request Database*.

One might expect that a request is submitted directly to the *jBPM process engine*, but each request is put into a database to ensure that no single customer request is every lost due to the process engine begin unavailable. This is required by a banking regulation that ensures that no risks are taken with customer submitted information. We must and will always be able to trace and audit every single step in the chain of events from customer request to product delivery. This small design step has been left out of the component diagram as it happens underwater and is of little importance to industries where intensive risk protection is not needed; we mention this in the interest of completeness.

uman tasks. A human action interface was implemented to provide functional administrators with the ability to deal with tasks as they drop out of the automated process for various reasons. Furthermore, Service Center employees provide input to the system through another interface with the document monitoring section of the process flow. Communication with the customer can require for a human task to be performed, such as customer's reply to questions which needs to be judged on completeness, correctness, and validity. This input to the jBPM process flow causes pending processes to be triggered into their next stages, to be stopped, or to be restarted. The interfaces have been created in-house by the project development team.

Within the project process definition it is always possible to encounter problems, planned or not, that need human intervention to solve. This intervention is called a human task, where the process is dumped into a task bucket for further action by an authorized person. We refer to the need to invoke human tasks as having the process *fall out* of the process flow. This fall out can then classified as either technical or functional. The first is often related to some error in processing a request within a process step, the latter is related to a problem in the application flow logic. When we look at full scale STP we are concerned with processes that by definition contain planned functional fall out points in their process descriptions.

STP Purchasing supplies a web based Java interface that provides an interface for humans to manipulate the tasks that they have been authorized to view. This component makes use of web services in the SOA layer to retrieve and manipulate process data located in various locations. It is mostly concerned with the *Request Database* where we find the complete request data structure that is maintained during the process life-cycle. One example of a functional fall out is a planned review of the applying customers credit rating results. This process might legally require that more than one person must review the customer's rating results before approving them as new bank customers.

Rule engine. This is a non-FOSS component supplied by a third party which we access from STP with BPM projects for business rules. This allows the business entity to maintain their own rule set regarding their businesses unit within the financial organization. For example, within a savings product you will have various rules and regulations as to the various conditions that must be met before a customer can be allowed to purchase that specific product. These rules and regulations can change over time or due to a special offer on that product during a specific time frame. It is often a wish from the contracting business unit to be able to manipulate these rules and regulations without having to contact the software vendor (i.e. project team).

JBOSS: jBPM and Service Layer (SOA). The application server is an open source component called JBOSS [11], from the JBOSS component family we have adopted the jBPM engine [12] and its process definition language (PDL) implemented in jPDL [13]. These are the main FOSS components in our project solution and are considered core components in the enterprise architecture.

The jBPM process engine is used for all BPM projects, so component selection was not an issue. The BPM process flows are defined by the information analyst together with the business customer for the application. It is a process involving workshops and use cases. It provides the lead developer of the project with a starting point, in the form of a process flow. This is mapped almost one to one into the process definition language which delivers a jPDL file. The resulting process definition is used for matching nodes to business services. In most cases this is a one to one mapping and the design of the services is the most time consuming part of the implementation. Should there be any technical details that call for adjustment to the flow, consultation ensues with the information analyst, and eventually with the business customer. Individual developers are then given technical designs based on use case realizations that allow them to integrate their implementations into the proper process steps.

The project was completed using only simple nodes that contain all business logic in plain Java. Basic service calls were combined in the Java code to achieve what later could be implemented as a complexer business service. There were no nodes implemented as actual wait states, where the process can wait for action from an external system. Our backend systems are not yet setup to trigger jBPM process instances to allow for real wait states. To facilitate wait states, a polling mechanism was used at points in the process were external systems needed to be checked for completion of a task. For example, while waiting for a customer to correctly identify themselves by returning a signed contract with a copy of a valid identification, the process will use a scheduler to periodically poll the backend system via a web service to determine if the identification has been completed. Once completion is detected, the scheduler triggers the process via a web service. Furthermore, there are the standard decision nodes, transitions, and human task nodes within the project's process implementation.

We have implemented a standard Service Oriented Architecture (SOA) [14], referred to in-house as our Service Oriented Architecture Layer (SOAL). Granularities of the services in this layer have been defined as basic services, business services, and some very simple composite business services (CBS) [15]. A basic service brings the existing transaction out of the backend system and makes it available through a web service. For example, to validate a postcode, the basic service *postcodeCheck* has been created to expose the backend mainframe transaction that checks if a given postcode is valid. The business services concern complex processing that may consist of one or more basic services. One of the more complicated issues is that of allowing the existence of CBSs in our SOA layer. These are business services that can contain not only calls to basic services, but to other business services as well, if the business service being called is in the same classification category as the caller.

The SOA layer deploys web services with versions. If a new release of the SOA layer contains services with interface changes, then the version of the release will be increased. To support backwards compatibility, a total of three versions is maintained for production applications to use. This allows for applications to upgrade to the newer versions over time.

Back-end systems. These systems can be anything in the wide variety that exist within our banking infrastructure: banking applications that provide and interface, external third party services, legacy systems, or some form of data storage like a data warehousing solution. It should be noted that these systems are always approached from our projects via the SOA layer in the form of a web service. We will provide the three most important backend systems that are used in STP Purchasing.

A *request database* was implemented for tracking each purchasing request as it migrated through the BPM process flow. This was the direct implementation of our purchasing request data model. As stated in context of the *web interface* and *human task* components, this database is filled with the initial request data, manipulated by the process as it migrates through the various steps, and directly affected when technical or functional fall out occurs. Access is arranged by a very specific service, dedicated to accessing, reporting, and updating data in the database. I works for the web interface, the human task interface, and from inside the process itself.

Another important component in the backend is the *customer information system*, used to maintain all customer and prospect contact information. This is a marketing data pool and there is a specific service dedicated to accessing and updating the information kept here.

A central system in our backend network is a legacy COBOL mainframe. This is is where the bank customers are managed and it is accessed via web services that make use of a Java communication layer. This layer bridges the gap between Java and COBOL mainframe functions which are provided when functionality is exposed from the mainframe.

3.3   Customer Testing

From the very beginning of the project, customer input was sought. An initial prototype was created, four customers and four internal customer support personnel were invited to conduct usability testing in a controlled environment. These 8 sessions where 90 minutes long, each dealing with a single respondent and a task assignment walk through. The walk through was done by the respondent with verbal communication accompanying all actions which were recorded by an observer sitting in a different room with a hidden view.

Even though it was a small usability test, it did provide relevant details which lead to advice for the development team in the areas of information structure, interaction, navigation, content, graphical information, style, layout, and features. Our view is that any steps taken to improve the customer satisfaction should be exploited to the fullest.

Another point in customer testing occurred before the project was released into production. It is was a last test that the business users took to test out the entire project. The testing users were guided by a test leader during the earlier project iterations to develop functional stories. These were then set up in the databases to allow them to test actions on submitting new requests, handling functional fall out, schedulers, and other such actions as they deemed necessary for project acceptance. This is a standard practice in our project release cycles and it remains a valuable feedback loop for finding functional problems before the project hits production status.

3.4   The Running Process

Empirical data providing results concerning running STP Purchasing in production since February 2008 will be presented below in an overview. The numbers represent the total number of processes per month, with a rather large spike in the months starting in September 2008. This was the beginning of the world wide Financial Crisis, which lead many Dutch citizens to spread their savings to different financial institutions.

Taking a look at Table 1 we can clarify some of the dips and peaks in the numbers. In February 2008 the project was released half way through the month, resulting in a low start number. It picked up steam and was pretty steady until August 2008, which we believe is due to the vacation period when most Dutch people tend to be on their holidays and away from computers. In September we

Table 1. Production process overview - 2008 monthly

Month	Requests
Feb	750
Mar	2750
Apr	2000
May	1200
Jun	1100
Jul	1500
Aug	850
Sep	4250
Oct	2250
Nov	1000
Dec	1800

Table 2. Status overview of customer processes

Status	Percentage
Completed on time	52%
Rejected for various reasons	8%
Human action (functional)	0.7%
Human action (technical)	0.3%
Currently in a fall out status	4%
In Document Monitoring	12%
Taken out of STP flow, completed by hand	23%

see the explosion of interest due to the Financial Crisis, followed by a leveling of interest. At the end of November 2008 the second set of 5 *deposito* products hit production. Unfortunately, we cannot say much about their impact as the December 2008 results are again of a partial month.

Another view of results is given in Table 2, which shows us percentages of the various statuses a process can be in. We must take into consideration that our metrics are limited and that we are only able to report on the process totals. Even so, it is encouraging that the amount of functional and technical fall out that needs attention are both less than 1% of the total. Also encouraging is that over 50% of all processes are completing on time. The ones that do not complete on time and are listed in *Document Monitoring* tend to be waiting for customer reactions to documentation problems as previously discussed. We have a timer running that ensures a customer receives reminders several times. Should the customer not reply at all, we eventually abort the request. The category listing 23% of processes taken out of the engine and completed by hand needs more explanation. This feature was added to allow special cases to be handled in the original manner, by hand.

With only 8% being rejected due to various reasons, it appears we are hitting the target audience and providing a process that is effective.

4   Observations

Not everything is as pretty as it seems, with issues remaining for both the business and technical sides of the playing field. We take a closer look at these, starting with the business challenges.

4.1   Business Challenges

Even though the use of business process models is proving itself successful at SNS Bank, there is room for improvement concerning the activities of conceptualization, communication, and engineering that are part of the ongoing development process.

Quality of business process models is a notion that has many aspects and thus is quite complex [16]. Engineering-oriented, mathematics-based aspects are involved (correctness, formal expressiveness, and various more specialized aspects such as mentioned by Vanderfeesten et al. [17]), but also social aspects (validation, agreement through collaboration, and common understanding [18]). For high quality process models to be realized, sufficient investment in detailed knowledge exchange and discussion is required.

The main challenge is a common one: the business (analysts, managers) have the best knowledge of detailed processes to be supported/automated, and have the authority to decide about them, yet are neither willing nor able to be too actively and intensively involved in high-quality, detailed, engineering-like specification of business processes, which they consider a "technical" job. Technicians, on the other hand, resent being forced to guess at details required for successful implementation and point out that "technical" is not the same as "involving detailed, precise, and well-conceived descriptions". This is mostly a cultural issue [18], but therefore also a deeply rooted one.

In an ideal situation, we would still need the people with the proper knowledge and authority concerning the business to describe and/or design products and processes. How this can be realistically achieved in the short run is an open question yet. Possible options include:

- Teach business people to read (at least) and create (less likely) formal modeling techniques.
- Find and use alternative ways to represent formal process models; verbalization, perhaps, or alternative (simpler) schemes.
- Encourage and allow business-oriented stakeholders to get involved in more detailed process modeling.
- Arrange for discussion and negotiation about process models to be optimally collaborative from the start, i.e. involve all relevant stakeholders at an early point and create explicit agreement (e.g. in workshops). The more divergence occurs in this phase, the more the diverging parties will fight for the survival of their initial ideas later on, and the harder it will be to reconcile alternative models. Discussion should take place, certainly, but not because effort and authority has been invested in particular diverging positions.

4.2   Technical Challenges

There are some interesting technical challenges that need to be watched for future projects. They cover issues concerning BPM, business logic, and (business) service releases. A currently completing BPM reference implementation [19] project has taken a closer look at these challenges and has come up with a few solutions and suggested ways of dealing with them.

Starting with the BPM issues, we have spent much effort to move the business logic out of the BPM process engine and down into the architecture to the SOA layer. This keeps the BPM engine lean and mean, requiring a lot less testing during the deployment phases of a project. Once the BPM flow is working, tests are passing, handlers call the correct services, and the infrastructure to support all of this is available, then there is not really much looking back. The main focus in searching out application problems are contained in the SOA layer. Developers spend their time testing and maintaining the business logic in the services, where it belongs. The delivered BPM flow should be almost maintenance free.

Many of the problems that the developers encountered with BPM process definition designs as described by Brahe [3] were avoided in our process by keeping the process flow definition, creation, and modification out of the hands of the developers. Modeling took place at a higher level, with a smaller group containing information analysts, business representatives, and the lead developer. This process led to a completed BPM process definition, in the process definition language, but expression in that language happend only at the end of the modeling process. We would like to look more into *directly* generating actual BPM process designs close to the chosen process definition language, *together with the business*. This is something to be further examined in the future.

Individual developers were able to concentrate more on working out the individual process steps (nodes and handlers), the given initial business service designs, test coverage, and documentation. This has worked well for us and we will continue to use this approach in the future.

Although there has been some literature on the use of SOA [20,21], we have found that most of the issues discussed where of little help when dealing with our own service construction. It seems that issues are often related to local conditions and infrastructure limitations. One complex issue arose in our environment, the issue of unreliable services due to all web service calls being implemented over the HTTP protocol [22]. The problem is more complicated when the basic services, themselves mapping to single backend transactions, are unreliable. It is conceivable that a service call is made to some complex business service that makes use of several basic services, and that it fails somewhere in the processes of executing basic services. We have no ability to implement anything other than a functional rollback and often are not sure what state the backend systems are left in.

There are potential problems with any service releases in the SOA layer that migrate to a major version number. For example, all minor version number releases from v1.0 to v1.1 of a given service contain no interface changes. These are therefore backwards compatible and should continue to work with all previously written consumers of the service. For major version changes, such as v1.1 to v2.0, we are confronted with a service containing an interface change that might break existing consumers of that service.

Service granularity has started to become a problem with more and more projects attempting to make use of basic, business, and composit business services that they find in the SOA layer. We hope to spend more time on looking into composite business service issues and lay some ground work with regards to guidelines for future projects.

A very sticky problem that has raised its ugly head is what to do with BPM process instances that are running when the new service release is planned. We are looking at our options at this time but have come up with the following strategy to provide a choice depending on the given situation:

1. Phase out older service versions when all old process instances have completed.
2. Build service converters that translate calls between different versions.
3. Activate a new BPM process instance for each existing old process instance.
4. Build a process converter that translates old processes into the new process definition (one time).
5. Human interaction to guide the process or complete the process flow.

This is an integral part of our current SOA service release strategy and can be found in the internal SOA documentation.

A solution is currently being tested that provides a custom class loader for each individual jBPM process engine. This allows each deployed process definition to provide the exact service version for each service it uses. This allows different deployed processes to access any of the SOA layer deployed service versions, independent of each other. This will have a positive effect on testing phases when multiple processes can be deployed on a single jBPM process engine, thereby saving extra hardware resources. This solution will also allow older instances of a process to be run next newer ones so that they can be phased out as mentioned above.

All contact from the process with internal systems is realized via web services. These calls are synchronous, but many of the backend systems are not. Many run batches which means that the web services provide transactions to functionality that can only report that the request has been received correctly. For example, a fictitious account is opened via a web service call, but this actually happens in a night batch run on the backend mainframe. The web service call will get the mainframe reply, *Account Opened*, but this process will not be actually completed until later. This indirectly means that web services can not be transactional or atomic in nature and a great effort is made in business service implementations to create as much of a functional roll back as can be achieved. More often than not, it means having to fall out of the process with a technical problem to be fixed by human hands.

At the time of this writing, a *state-proxy* is being implemented to allow for real wait states in the process definitions. When using a wait state, the business service call is done through our state-proxy. The process is then put into a wait state and the proxy handles the web service call, returning either an exception or the results. This state-proxy can then be expanded with extra plug-in like functions, such as dealing with service windows for known down time on backend systems running a batch, allowing for technical retires to services that can be offline for short periods of time, and dealing with standard exceptions. These plug-ins are on the drawing board for future implementation.

The scheduler discussed in section 3.2 is a point of concern. This does not scale well and in the future we will need to look into getting our backend systems to trigger on certain events. This should be possible and the discussion is underway.

Another nice-to-have would be to remove the non-FOSS rule engine discussed in section 3.2. We want to spend some time looking into the JBOSS rule engine in the coming year which seems to provide a solution that is integrated in our existing development tooling.

5   The Bene ts

As we have seen, the benefits of BPM are promising, based on the empirical data collected in the deployed production process. A closer look at the customer and development benefits will make it clear that there is much gained already.

5.1   Improving the Customer Experience

A key concept in the vision of this solution is that the customer must be central to the process. A customer centric business model is not new [9], but we feel that aligning the entire strategy to empower one's customers is breaking the mold.

As strategic products are made available through full scale STP with BPM we are able to adjust easily to customer needs. Products and product lines can be introduced into existing business processes in a cost effective manner. The flexibility to combine extends beyond products, product lines, and selling channels to become a very effective tool to reach customer bases in a timely and personalized fashion.

Customer communication can be personalized and tailored to specific processes, products, and customers' personal needs as the data generated by their behavior within the processes is documented. There have been very positive reactions from customers with regards to the speed, quality, and the level of detail in communications.

5.2   Development Process Improvements

The initial STP Purchasing project has provided a starting point for the IT department to build on for future full scale STP with BPM projects. Lessons learned and best practices are being applied, resulting in some interesting improvements to the process.

To our initial surprise, BPM process definitions can be easily changed with a minimal impact on the development time. The work is not in the process definition, but in the business services and basic services in the underlying structure. A standard way of implementing process nodes and testing has made this part of the development process much less critical. It is important to focus on what we call the *Happy Flow* during initial development. This is the backbone of the process flow which represents a positive test case that processes as expected. For example, we would focus in the STP Purchasing project on a single saving product being requested by a verified and known customer of the bank. This means that you do not have to deal with any exceptions during the initial run through your process implementation. The focus of the first iteration of development is to get this *Happy Flow* working. By providing a quick working Happy Flow, the business can be shown tangible progress in the project at an early stage.

With an ever growing base of BPM process definitions it is clear that the time to market for similar products is much quicker. We have projects with estimates ranging from one third to one half of the initial development hours put into STP Purchasing. This is quite a big improvement. One thing of note here would be that the development of business services should always be carefully considered, as they tend to be the focus point of complexity.

The initial process definitions as provided by the information analysts and business analysts are not in our process definition language. Much depends on the quality of this process flow model, but with some care and attention to this step it is not too much trouble to map this process flow model to our process definition language. The generated image of the flow is a very good communication tool with the business. No better way to let them see the business services and understand where the development time is spent. Bringing the business closer to the development team with regards to communication over the process flow has been a positive experience that we would like to see continued.

6   Future Plans

In the coming year(s) we will be tackling the projects described below. They are a clear sign that the success that has been achieved with previous full scale STP with BPM projects is relevant. SNS bank is moving ahead full steam with projects for certificates of deposit, debit accounts, and the migration of existing internal and external service processes.

Since STP Purchasing completed in the beginning of 2008, work has started on expanding the product category with five new *deposito* products. This project went live in November of 2008, just in time to provide the Dutch market with an easy way to spread their saving money around between banks: note the peaking numbers in the empirical data that reflect the public reaction the the start of the Financial Crisis in September 2008.

6.1   STP Payments

This project will take place in 2009, with plans to focus on debit accounts for standard customers, children, and students. The solution will need to implement the following business processes for these accounts:

– requesting and receiving a bank pas for the account
– requesting and receiving internet access for the account
– requesting and updating the credit limit for the account
– requesting and receiving a new credit card

6.2   Migration Service Processes

A service process migration project is looking at migrating 169 internal processes to make full use of BPM and STP. These are processes that internal SNS Bank employees make use of to process various customer needs. All of these processes will make use of our full scale STP with BPM solutions as much as possible. In 2008 one of these service processes was put into production: a process to allow customers to submit a name or address change through full scale STP with BPM.

A list of items that have caught our attention and imagination are provided here without further discussion. These are possible points for exploration that might provide added value to the process of deploying full scale STP with BPM in the future.

- enable business customer to be more involved with "business engineering"
- better motivate business customer to be involved with "business engineering"
- enable business customer / information analyst to deliver better PDL models as input to the development process
- enable migration of existing process definitions to new process definition releases (results of reference project)

7   Moving Ahead

In this paper we presented the efforts of a Dutch bank at migrating from traditional banking services to a full scale STP with BPM driven financial institution. The components being used to realize the STP Purchasing project were described and the resulting empirical data were presented for evaluation. The issues and benefits were covered along with the challenges yet faced by both business and IT development. The large shift in strategy has started to deliver the desired results and these will continue to roll in as future full scale STP with BPM projects are implemented.

The positive effects on customer interaction, improvements on accelerating product deployment, and more flexible product/customer support channels have energized some internal ideas about becoming a facilitator to external third party enterprises. Imagine a future where individual entrepreneurs would be able to open a banking store with complete full scale STP with BPM selling channels for products and services.

We hope that our experiences, lessons, and observations will be of value to the industry as a whole. This is a financial industry story, but it could be applied to many different situations and time taken to learn from this story would be well spent.

References

1. Khanna, A.: Straight Through Processing For Financial Services: The Complete Guide. Academic Press, Burlington (2007)
2. van der Aalst, W.M.P., ter Hofstede, A.H.M., Weske, M.: Business Process Management: A Survey. In: van der Aalst, W.M.P., ter Hofstede, A.H.M., Weske, M. (eds.) BPM 2003. LNCS, vol. 2678, pp. 1–12. Springer, Heidelberg (2003)
3. Brahe, S.: BPM on Top of SOA: Experiences from the Financial Industry. In: Alonso, G., Dadam, P., Rosemann, M. (eds.) BPM 2007. LNCS, vol. 4714, pp. 96–111. Springer, Heidelberg (2007)
4. Guerra, A.: Bloomberg Aims To Simplify Straight-Through Processing. On: InformationWeek (December 18, 2000),
http://www.informationweek.com/817/bloomberg.htm

5. Strickland, R., Aach, D.: Getting to straight-through processing: in theory, there is a way to deliver faster and better service in the mortgage lending business. On: BNet Business Network (February 2006),
   http://findarticles.com/p/articles/mi_hb5246/is_/ai_n29277448
6. The Free Dictionary (February 10, 2009),
   http://encyclopedia2.thefreedictionary.com/Straight+Through+Processing
7. Answers.com (February 10, 2009),
   http://www.answers.com/topic/straight-through-processing
8. Investopedia (February 10, 2009),
   http://www.investopedia.com/terms/s/straightthroughprocessing.asp
9. Heckl, D., Moormann, J.: Matching Customer Process with Business Processes of Banks: The Example of Small and Medium-Sized Enterprises as Bank Customers. In: Alonso, G., Dadam, P., Rosemann, M. (eds.) BPM 2007. LNCS, vol. 4714, pp. 112–124. Springer, Heidelberg (2007)
10. Java Technology (March 18, 2008),
    http://java.sun.com (March 19, 2008)
11. Jboss.org: Community Driven (March 19, 2008),
    http://labs.jboss.com
12. jBPM Overview (March 19, 2008),
    http://labs.jboss.com/jbossjbpm/jbpm_overview
13. Welcome to jBPM jPDL (March 19, 2008),
    http://labs.jboss.com/jbossjbpm/jpdl
14. Erl, T.: Service Oriented Architecture: Concepts, Technology and Design. Prentice-Hall, Englewood Cliffs (2005)
15. Neuman, S.: Composite Business Services. IBM Global Business Services (October 25, 2008),
    http://www-935.ibm.com/services/us/index.wss/offering/gbs/a1027243
16. Recker, J.: Towards an Understanding of Process Model Quality. Methodological Considerations. In: Ljungberg, J., Andersson, M. (eds.) Proceedings 14th European Conference on Information Systems, Goeteborg, Sweden (2006)
17. Vanderfeesten, I.T.P., Cardoso, J., Mendling, J., Reijers, H.A., van der Aalst, W.M.P.: Quality Metrics for Business Process Models. In: Fischer, L. (ed.) BPM and Workflow handbook 2007, pp. 179–190. Future Strategies Inc., Mississauga (2007)
18. Hoppenbrouwers, S.J.B.A.: Community-based ICT development as a multi-player game. In: Conference proceedings of What is an Organization? Materiality, Agency and Discourse, May 2008, University of Montreal, Canada (2008)
19. Schabell, E., Benckhuizen, J.: Software Architecture Document – jBPM Reference Project. SNS Bank IT, s-Hertogenbosch (2008)
20. Mahajan, R.: SOA and the Enterprise – Lessons from the City. In: IEEE International Conference on Web Services (ICWS 2006), pp. 939–944. IEEE Computer Society, Los Alamitos (2006)
21. Acharya, M., Kulkarni, A., Kuppili, R., Mani, R., More, N., Narayanan, S., Patel, P., Schuelke, K.W., Subramanian, S.N.: SOA in the Real World – Experiences. In: Benatallah, B., Casati, F., Traverso, P. (eds.) ICSOC 2005. LNCS, vol. 3826, pp. 437–449. Springer, Heidelberg (2005)
22. HTTP – Hypertext Transfer Protocol (February 27, 2008),
    http://www.w3.org/Protocols (March 19, 2008)

Progressing an Organizational Approach to BPM: Integrating Experience from Industry and Research

Tonia de Bruin[1] and Gaby Doebeli[2]

[1] Queensland University of Technology, Department of Science and Technology, Brisbane, Australia
t.debruin@qut.edu.au
[2] Griffith University, Department of Information Systems, Brisbane, Australia
gaby.doebeli@qr.com.au

Abstract. In 2002, Company Q knew it had a problem. No longer could it continue to run its operations as it had previously. Disparate projects were having a counteractive effect. Changing legislation and regulations were increasing reporting requirements and competition. Increased usage of its transport networks were resulting in scheduling difficulties, delays and customer dissatisfaction. A thorough review of alternative business management approaches indicated merit in adopting Business Process Management (BPM) as an organizational approach. At the time however, the process of how to adopt such an approach had received little attention in either academic or practitioner literature. Consequently, Company Q approached QUT for assistance with progressing and measuring BPM as a holistic approach to managing an organization. This paper reflects upon the role of the study in Company Q's subsequent BPM journey.

Keywords: Business Process Management, Organizational Transformation.

1 Introduction

Over recent time, there has been growing interest in improving and managing an organization's processes. From 2005 to 2009, [1] has found that process improvement is the number one business priority of CIO's. Furthermore, the importance of BPM is recognized by [2] who found that 97% of European organizations surveyed considered BPM important to the organization and only 3% had not commenced BPM practices. Similarly, in an earlier study, [3] found that 96% of respondents were engaged in 'some form of process management' with formal programs adopted by 68% of these respondents. A number of drivers are contributing to the continuing interest in adopting a process approach including:

1. Need to improve responsiveness and quality and, to manage competitive threats [2]
2. Globalization, changing technology, regulation, the action of stakeholders and the eroding of business boundaries [4]
3. Competitiveness of industry within the international marketplace [3]

E. Proper, F. Harmsen, and J.L.G. Dietz (Eds.): PRET 2009, LNBIP 28, pp. 34–49, 2009.

According to [5] and [6], a process approach to business increases competitive advantage by reducing cycle times, utilizing new information technologies and obtaining managerial control. More generally, [6] and [7] suggest that a process focus helps to achieve higher (sustainable) performance with strategies including reducing costs, resources and/or overheads.

Another benefit, according to [2], [6] and [7], is that adopting a process approach leads to increases in customer satisfaction and an improved ability to respond to customer needs. It does this by improving an organization's focus on the customer [6], [8], and introducing greater flexibility [7]. Other suggested benefits relating to customers include reducing time to market and improving service delivery [5] and [6], and improving quality [3], [6] and [7].

Furthermore [2], [8] and [9], indicate that a process approach has a positive impact on change management and cultural issues. In a similar vein, [10] indicates that BPM practices can lead to a reduction in turf mentality and [6] suggests that improved teamwork is possible. Similarly, [8] indicates that increasing the level of employee empowerment leads to a reduction in cross-functional barriers.

Despite high interest, strong drivers and the recognized benefits associated with adopting a process approach, organizations such as Company Q face many difficulties in adopting and progressing BPM. For example, [2] indicates that despite 97% of organizations expressing a high interest in BPM as few as 27% of those surveyed were at more than a basic level of adoption, suggesting a disconnect between interest and progression.

Furthermore, [11] raises a concern that the latest round of process thinking may become yet another management fad with continuing evidence of unsuccessful attempts whereby organizations fail to reap the benefits of their efforts. This concern is not aided by the inconsistent use of terminology in the BPM domain, an example of which is the use of the term BPM itself. In extant literature, this term has three common interpretations including:

(1) BPM as a software technology solution (cf. [12] and [13]), although the terms of BPMS or PAIS are now becoming more commonplace (cf. [14])
(2) BPM as an approach to the management of a process/s within an organization, through the various stages of the process lifecycle (i.e. a lifecycle approach) (cf. [3], [5], [6] and [15])
(3) BPM as an approach to the management of an organization, seeking to create a process-oriented organization, as opposed to a functionally-oriented organization (i.e. an organizational approach) (cf. [2] and [18])

From a theoretical perspective, [16] and [17] discuss shortcomings of theory arising from interpretations (1) and (2), with regard to adopting an organizational BPM approach that captures the richness of organizational context and the temporal aspects of progression.

Similarly, in practice, the use of the term BPM is not always clear or consistent. Anecdotal evidence suggests that some organizations prefer not to use the term due to past problems with the adoption of BPR initiatives, instead using terms such as Business Transformation and Organizational Renewal Program. Furthermore, in practice, BPM Initiatives do not align neatly with the interpretations of BPM indicating they are not mutually exclusive. For example, in adopting a lifecycle or an organizational approach a company may also adopt a BPMS or a PAIS. Similarly, in adopting an organizational approach, a company may also adopt a lifecycle approach to managing the processes within the organization[1]. These conflicting views potentially affect the expectations of practitioners seeking to adopt and practice BPM.

This paper presents aspects of a research study that commenced in 2004 and aims to bridge the gap between industry and research with regard to progressing BPM within organizations. The structure of this paper is as follows. Section 2 of this paper provides the background of an industry case study, Company Q. Section 3 summarizes aspects of the study undertaken by researchers with a view to measuring the progress of BPM Initiatives within organizations. Section 4 details Company Q's application of a key outcome from the study, the BPM Capability Framework. Section 5 details the researcher's subsequent application of the BPM Capability Framework within Company Q, in order to progress the development of theory regarding BPM as an organizational approach. Section 6 concludes this paper with a review of the key points.

2   Background to Company Q

Company Q is one of Australia's largest and most modern integrated transport providers. Operating on a transport network of more than 10,000 kilometers across the continent, its services include Passenger Service, Freight Service and Network Access Provision. Company Q has annual revenue in excess of *A3.5 billion and manages assets of *10 billion. Company Q is among the nation's longest running service enterprises with approximately 15,000 employees throughout the country. Company Q is a Government Owned Corporation (GOC) directed by a Board that is accountable to two ministers within the Queensland State Government. In 1999, a move to increase the commercialization of some State Government operations resulted in Company Q moving from a monopoly government provider to becoming a national commercial operation in a competitive business environment. This change resulted in major challenges for the operation of Company Q.

As a consequence of the move to commercialization, disparate projects were having a counteractive effect. Changing legislation and regulations were increasing reporting requirements and competition. Increased usage of its transport networks were resulting in scheduling difficulties, delays and customer dissatisfaction. Attempts to improve operations by applying methods like Quality Assurance (QA), Total Quality Management (TQM), Business Process Reengineering (BPR) and Business Process Improvement (BPI) had resulted in limited success. Consequently Company Q was seeking to adopt BPM as a holistic approach to the management of the organization.

2.1   Starting BPM within Company Q

Against this backdrop, in 2002 Company Q's Board and Senior Executives assigned the Chief Strategy Officer (CSO) to lead a major change program to establish a sound platform to achieve service excellence and allow further growth of the business. The

---

[1] In recognition of these potential variants, this paper uses the term BPM Initiative to refer to BPM program of work within an organization.

overall objectives were to (1) gain transparency of processes and cost, (2) achieve accountability throughout the different levels of management and (3) operate as a successful organization that makes profit. The CSO established three program streams called Performance through Governance, Performance through Business and Performance through People. The program stream of *Performance through Business* included a project that was to investigate *Business Process and Systems*.

The first phase of the *Business Process and Systems* project led to the identification of an enterprise-wide BPM approach as a means of addressing some of the operational and strategic issues facing the organization. However, getting support for adopting such a BPM approach and developing the initial frameworks was difficult due to (1) conflicting literature and practice regarding what constituted an enterprise-wide BPM approach and (2) a lack of guidance as to how to go about adopting such an enterprise-wide approach.

The second phase of the project included making the frameworks operational in order to embed BPM principles and practices within the organization. The CSO established a BPM team led by the Business Process Design Adviser (BPDA)[2]. The BPDA reported directly to the CSO. In the first instance, the BPDA was responsible for the establishment of the methods and techniques within the framework, and the introduction of these to the organization.

Due to the failings of past endeavors arising from the implementation and use of methods including TQM and BPR, and following an extensive review of literature and practices within other organizations, the BPDA affirmed the earlier belief that a BPM approach focused on the management of the organization was appropriate to addressing Company Q's needs. However, subsequent investigation revealed a lack of a suitable means by which to (1) understand existing practices and to gain guidance on progressing and embedding BPM practices within the organization and (2) an inability to measure the progression of BPM practices adopted within the organization.

3 Progressing and Measuring BPM within Organizations

To address these issues, Company Q approached QUT for assistance, resulting in a study into the progression and measurement of BPM Initiatives within organizations. In defining the study, the researchers recognized an inherent tension between the evolutionary nature of progression and the static notion of measurement at a given point. Consequently, the researchers distinguished between these concepts using the terms of *BPM Progression* and *BPM Maturity*. BPM Progression referred to the dynamic and evolutionary journey of a BPM Initiative and BPM Maturity referred to the static measurement of progression from time to time. Fig 1 shows the basic premise of the study, including the theorized relationship with BPM Success and Process Success.

To progress the study, the Researchers developed an initial conceptual model encompassing factors that were seemingly important when adopting a BPM approach[3], identified from extant literature. Details on the development of the early conceptual model are in [19] and [20]. During 2004, Company Q participated in a case study

---

[2] The BPDA is co-author of this paper.
[3] This initial conceptual model had its base in all three recognised BPM approaches being the technical, life-cycle and organizational approaches.

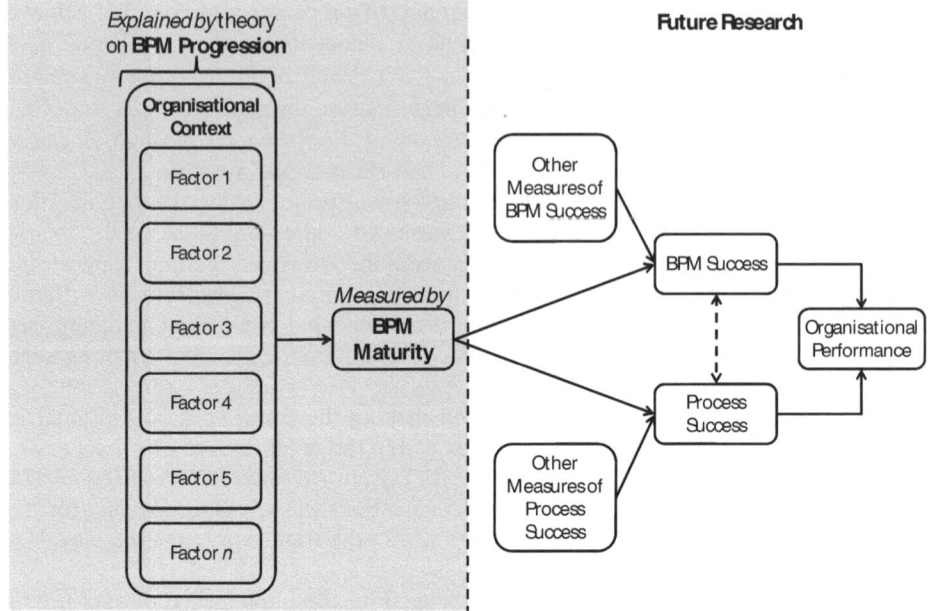

Fig. 1. Positioning the BPM Progression and Measurement Research

conducted by the Principal Researcher to refine the conceptual BPM model and to pilot early versions of the survey instruments for measuring BPM Maturity, within three of its lines of business. Details of the conduct of this case study, including the key outcomes and subsequent changes to the model are available in [21]. However, to provide context to this paper, the next sections detail the key findings for Company Q, together with a summary of selected practical and theoretical consequences arising from the case study.

3.1 Key Findings for Company Q from Application of Model

From Company Q's perspective, the key findings from the 2004 case study confirmed that they had focused most attention on selecting and implementing methods and tools for BPM at an operational level. This accurately reflected the early focus on the tools and methods of BPM (such as BPR and BPI) which Company Q had subsequently found to be insufficient to support the transformational change required.

The case study found that Company Q used business improvement methods in pockets of the organization, with the assistance of some information technology (IT) tools. However, there was not an overall common framework to guide the selection and use of methods and tools in place. The result being that, there was no consistency in the actual methods and tools applied throughout the organization. Nor was there any consistency in their manner of application across the various business units, even when people were using the same method / tools. For example, the organization had chosen standard modeling notations (i.e. BPMN and IDEF), however as Company Q had implemented no other governance around the notations, there was no consistency in the use of these notations within the organization.

Similarly, employees within Company Q made ad-hoc attempts to improve processes using the available tools and methods in projects. However, the projects they undertook often did not have strong alignment or linkage to the overall strategic objectives of the business but were in reaction to an operational pain point. There were no process strategies or planning in place that outlined the activities required to relate process performance to the desired improvement and sustained business performance.

In addition, employees across the business had received little guidance on how to improve and build their process and process improvement skills and knowledge. The organizational culture showed a strong, functionally based, silo mentality focusing on improving individual functions rather than the end-to-end process that delivered the services to the customer. To compound this, people were not accountable for their actions or the consequence of these actions on process outcomes.

3.2 Consequences of the Study for BPM within Company Q

Company Q found that a key benefit of participating in the case study was that the Researchers provided a detailed presentation and report into the review of BPM practices within the company. This provided Company Q with an independent view of practices that added credibility to internally held views, making them more acceptable to some members of the organization. The study identified that the strengths within Company Q's BPM initiative were mainly in the area of BPM Methods and Technology. However, it also confirmed that the areas of greatest weakness were in Strategic Alignment, Governance and Culture. Thus, using the case study findings as a basis,

Table 1. Early Strategies for Progressing Company Q's BPM Initiative

Factor	Strategies for Progression
Strategic Alignment	- Align process information with strategy and technology - Commence the development of an Enterprise Process Architecture
Governance	- Develop the overarching BPM Concept including governing principles, terms and conditions based on the BPM approach - Integrate governing principles with the new Business Model - Distinguish the new Business Model from the narrower interpretation arising from the SAP-R3 implementation - Develop and assign specific process responsibility
Methods	- Establish clarity around use of Six Sigma, BPR and Lean Manufacturing
Information Technology	- Identify business requirements for an enterprise architectural tool that supported the BPM view
People	- Develop case studies to capture BPM knowledge - Train employees in the use of selected methods
Culture	- Establish process forums to facilitate knowledge sharing throughout the organization - Establish a BP community consisting of representatives throughout the organization - Evolve the BPM Concept more broadly within the organization - Identify ways of changing the attitudes and behaviors within different sections of the organization

the BPM team at Company Q was able to develop strategies to improve their BPM practices, particularly in the areas of Strategic Alignment, Governance and Culture. Table 1 provides an example of the strategies enabled within Company Q, from participating in the case study.

3.3 Consequences for Research

In addition to consequences for Company Q, applying the early model and piloting the survey instruments in the case studies had implications for the Researchers. In particular, the case study highlighted (1) the role of contextual variables in the progression of BPM at an organizational level and (2) the need for greater clarity and additional granularity within the factors contained in the model. Consequently, the researchers extended the model through the conduct of an international series of Delphi Studies. Details of the design and conduct of the Delphi studies are included in [22]. However, a summary of the Delphi studies and their outcomes is included here to provide context to the application of the resultant extended model within Company Q.

The Principal Researcher[4] conducted the Delphi studies from February to September of 2005. There were six Delphi studies in all, one for each factor of the model. Each study included between 10 and 20 BPM experts from USA, Europe and Australia[5]. The aim of the studies was (1) to agree a definition of each factor and (2) to identify the major items whose measurement would indicate advancing maturity in the factor. The Principal Researcher subsequently referred to final measurement items identified as *capability areas*. During each round of the Delphi studies, a panel of

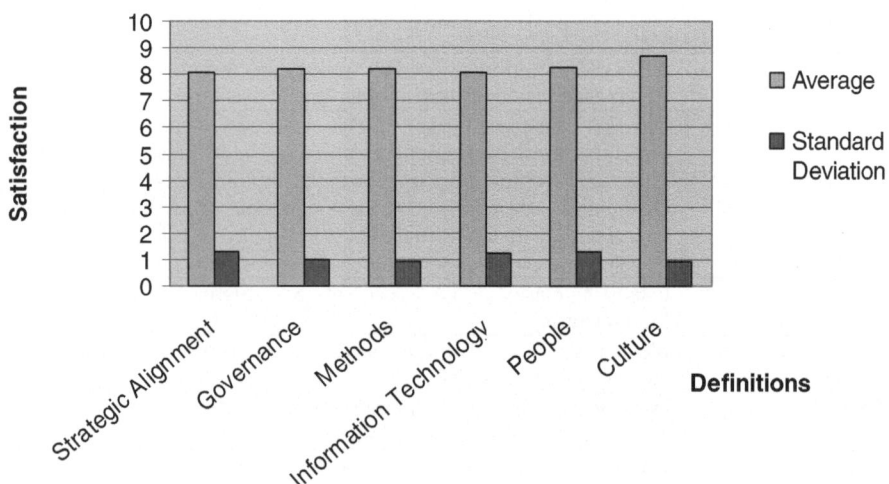

Fig. 2. Average Satisfaction and Standard Deviation of Responses for Final Factor Definitions

---

[4] The Principal Researcher is co-author of this paper.
[5] Representatives from Company Q did not participate in the Delphi Study series.

three academics[6] coded data for rating and comment by the Expert Panel in the subsequent round. Decision rules were set to assist in determining the end of each study. Fig. 2 provides a summary of the average satisfaction (with 10 being very satisfied) and the standard deviation for the final factor definitions.

Table 2 shows the final agreed definitions for each factor determined during the international Delphi Studies.

Table 2. Final Factor Definitions from the Delphi Studies

Factor	Definition
Strategic Alignment	...the continual tight linkage of organizational priorities and enterprise processes enabling achievement of business goals.
Governance	...establishes relevant and transparent accountability and decision making processes to align rewards and guide actions.
Methods	...the approaches and techniques that support and enable consistent process actions and outcomes.
Information Technology	...the software, hardware and information management systems that enable and support process activities.
People	...the individuals and groups who continually enhance and apply their process-related expertise and knowledge.
Culture	...the collective values and beliefs which shape process-related attitudes and behaviors.

In addition to the agreed definitions and the identification of the capability areas, during the Delphi studies, the Principal Researcher mapped all items and comments raised by the expert panel, to the final list of capability areas. Following the Delphi

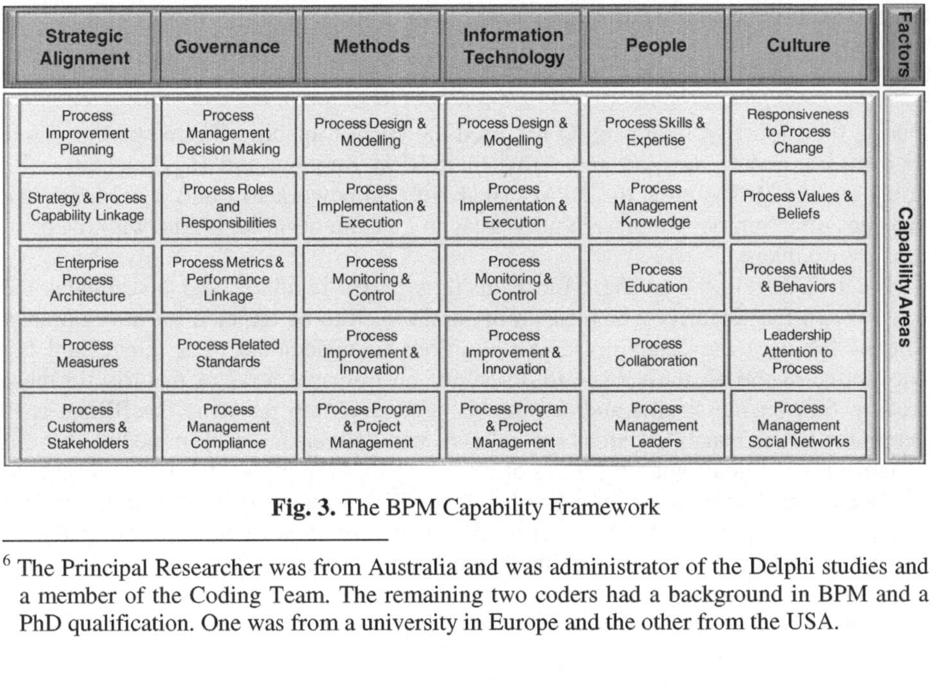

Strategic Alignment	Governance	Methods	Information Technology	People	Culture	Factors
Process Improvement Planning	Process Management Decision Making	Process Design & Modelling	Process Design & Modelling	Process Skills & Expertise	Responsiveness to Process Change	
Strategy & Process Capability Linkage	Process Roles and Responsibilities	Process Implementation & Execution	Process Implementation & Execution	Process Management Knowledge	Process Values & Beliefs	Capability Areas
Enterprise Process Architecture	Process Metrics & Performance Linkage	Process Monitoring & Control	Process Monitoring & Control	Process Education	Process Attitudes & Behaviors	
Process Measures	Process Related Standards	Process Improvement & Innovation	Process Improvement & Innovation	Process Collaboration	Leadership Attention to Process	
Process Customers & Stakeholders	Process Management Compliance	Process Program & Project Management	Process Program & Project Management	Process Management Leaders	Process Management Social Networks	

Fig. 3. The BPM Capability Framework

---

[6] The Principal Researcher was from Australia and was administrator of the Delphi studies and a member of the Coding Team. The remaining two coders had a background in BPM and a PhD qualification. One was from a university in Europe and the other from the USA.

Studies, the Principal Researcher used this mapping and insights from the earlier case studies to develop high-level definitions for each of the capability areas. These definitions appear in [23]. Fig. 3 presents the six factors of the initial conceptual model together with their final capability areas derived from the Delphi Studies in a so-called **BPM Capability Framework**[7].

4 Applying the BPM Capability Framework in Company Q

From a practical perspective, the BPM Capability Framework is a useful tool for assisting in the direction and progression of BPM Initiatives within organizations. This is evident in a number of organizations applying the BPM Capability Framework in the progression of their BPM Initiatives. By way of example, this paper presents the application of the BPM Capability Framework by Company Q which began in 2005.

Representatives from Company Q have developed their understanding of the BPM Capability Framework through a number of avenues. On the one hand, they have maintained an ongoing relationship with the Researchers through regular meetings and informal discussions. On the other hand, they have participated in BPM forums including the Australian BPM Roundtable, the Queensland BP Trends Chapter and the Australasian Process Days Conference where updates on the framework and insights from its application have been presented. In addition, they have read published articles arising from the study, over time.

The following sections discuss Company Q's use of the BPM Capability Framework to (1) develop a BPM Roadmap to guide the progression of their BPM journey, (2) adjust subsequent BPM strategy and communication to their changing organizational context and (3) to map their lessons learned and integrate key findings into the future direction of their BPM Initiative.

4.1 Developing a BPM Roadmap

During their journey, Company Q has used the BPM Capability Framework to guide their efforts and to develop and refine their BPM Strategy and Implementation. In doing so, the BPDA uses the BPM Capability Framework to build a roadmap that provides direction on which capability areas to give attention to, in line with the business environment.

The BPDA reflects on the critical business issues regularly and investigates the causes to further identify which capability areas need to be targeted for development. The BPDA uses the capability area definitions to understand their intent, and her knowledge of the business issue to determine an informal level of maturity in these areas within business units and/or projects. The next step taken by the BPDA is to determine which capability areas she believes will deliver the most immediate benefit to achieving the goals and objectives of Company Q. In doing so, the BPDA is able to allocate resources and develop capability that will optimize the benefit to the organization from adopting a BPM approach. The Principal Researcher is not directly involved in the determination or implementation of these strategies, however, the

---

[7] The depiction of the capability areas in this manner does not mean the areas are hierarchical.

Principal Researcher and the BPDA meet regularly to discuss or clarify issues regarding the intent and interpretation of the capability areas and possible strategies and their implications.

Since 2006, the BPM team's role is the delivery of BPM services to the business areas using an internal consultancy arrangement. The team uses the BPM Capability Framework to guide the conduct of its consulting engagements, including the subsequent recording and documentation of the engagements and their outcomes. Every consulting assignment is carefully scoped including an initial BPM capability assessment. This assessment identifies additional activities to be conducted to further improve BPM capability as part of the project delivery. Every consultancy engagement is finalized with a workshop where project results are reflected on as well as the progression of BPM capability in the particular business area. Every project team collates their findings and learnings in a case study format for publication on the BPM portal site to share with others. Therefore, each project contributes to the improvement of different capability areas within the BPM Capability Framework, in addition to delivering the desired business need. The project summaries are an effective tool to further consolidate and communicate the BPM progress to the rest of the organization and for the BPDA to further set strategies to enhance capability areas within the model.

4.2 Adjusting Strategies to Fit Changing Organizational Context

Despite the advances that Company Q has made, the progression of an organizational approach to BPM is not without issues. Recent changes within Company Q that have influenced the progression of BPM include (1) changes in Company Q's business model and (2) changes to the organizational structure. By using the BPM Capability Framework, Company Q has been able to adjust its BPM strategies in response to these changes as shown in the following examples.

Changing Business Model

In 2008, the appointment of new senior management in Company Q led to a change in its business model and strategic direction. Company Q was redesigned from a business model as an integrated transport provider to a multiple company business model. The objective of the new business model was to create accountability and to increase the flexibility and agility of Company Q, making it more competitive in the market place. To implement the new business model, changes to the Corporate Governance Framework were necessary. Consequently, Company Q revised their Corporate Governance Framework from a strongly rule-based to a principle-based focus in recognition that one size does not fit all and in order to empower management in the decision making process by giving them greater accountability for business outcomes. Under the new Corporate Governance Framework, Practice Leaders (i.e. function and process owners) were called upon to translate existing rule-based policies into Practice Principles.

Since the new governance framework was put in place, questions have arisen about its effectiveness. Consequently, the Company Secretary engaged the BPDA to assist in a review of the organizations new Corporate Governance Framework. The BPDA was able to use the BPM Capability Framework and the BPM Principles in this review. The review found that, despite the involvement of the Practice Leaders, the accountability structure was still based predominantly on functional demarcations and that not all Practice Leaders were identified and/or included. The review also found gaps in the decision making process and ineffective information flow within the core processes of Company Q. The findings of the BPDA were subsequently supported by an independent review of the Corporate Governance Framework by an external party.

Changing Organizational Structure

Since commencing its BPM journey in 2002, a number of organizational restructures have led to significant changes in the roles and responsibilities of the BPM team. At times these changes have affected the manner in which the team operates or is resourced, whilst at other times, these changes have affected the location of the BPM team within the organization.

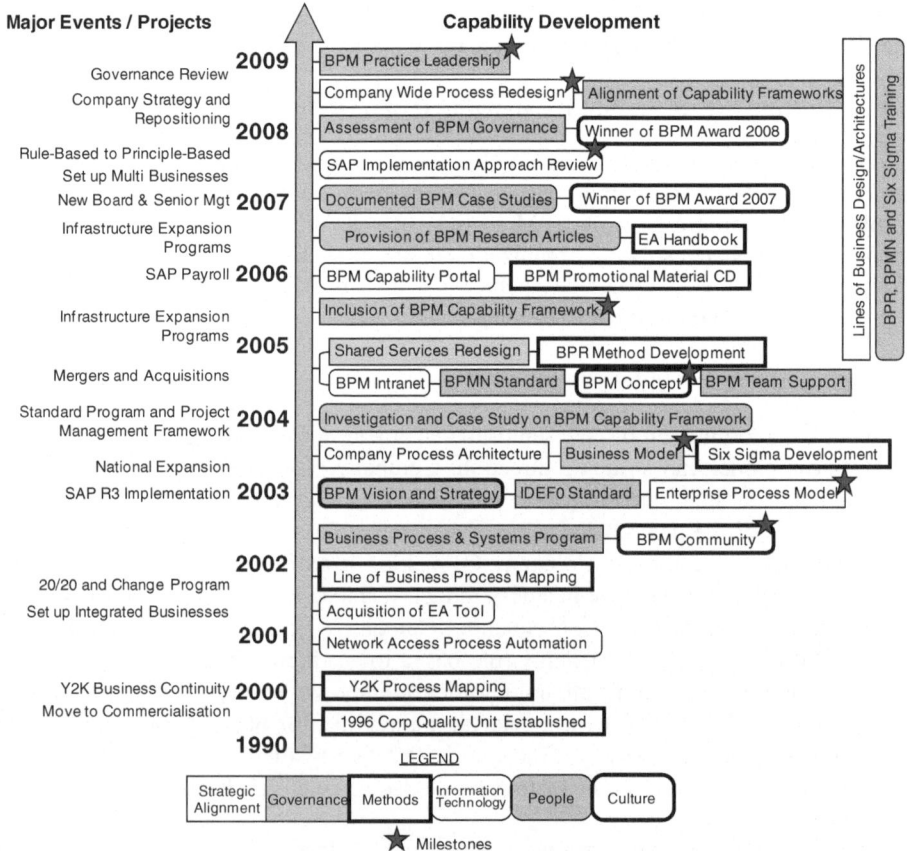

Fig. 4. Company Q's Major Events, Projects and Capability Development

In mid-2007, the BPM team was assigned their most recently defined role as the Practice Leader for BPM in the organization. The BPM professionals that make up the BP community in Company Q are now directed and supported by the BPM team. These professionals reside in the individual businesses, often within the Business Support areas, and work together with strategic planning, human resource, finance and IT functions as well as the areas of risk and project management. The recognition of the BPM team as a Practice Leader also means the team now works closely with other leadership teams of the organization to build BPM capability.

The use of the BPM Capability Framework throughout these structural changes has provided the BPM team with a stable framework independent of the changing organizational structure. This is evident in the redevelopment of the BP Community Portal, utilizing the BPM Capability Framework and underpinning the directory structure that

Table 3. Lessons on Progressing BPM

Factor	Lessons on Progressing BPM
Strategic Alignment	- Select projects that are of strategic importance and have senior management commitment
	- Strong connections between strategy formulation and selection of process improvement initiatives helps to optimize resource planning and allocation
	- Defining end-to-end processes and assigning ownership and accountability for process performance, including linking to individual performance measures, helps to optimize outcomes
Governance	- Putting BPM governance in early ensures clear direction and leadership and a common terminology
	- BPM Governance needs to be integrated into an overarching corporate governance framework
	- Process leaders need to be supported by their functional counterparts within an integrated governance framework to ensure optimal process decision making
	- Process related standards need to be developed throughout the journey as maturity increases in different areas
Methods	- A standard notion helps to provide consistent, reusable models and process information
	- The notation selected is not as important as its consistent application and ability to be supported by a suitable modeling tool
	- Multiple complimentary methods for process improvement are beneficial for matching the method to the purpose improvement project
	- Strong program and project management is needed to track the benefits for the organization from the improvement projects and the BPM program itself
Information Technology	- A common process modeling/repository tool is essential when progressing an enterprise wide BPM approach
	- Matching the tool to the purpose of the modeling becomes important over time
People	- Hands-on involvement in projects is an effective way of learning and embracing the BPM approach
Culture	- An organizational approach to BPM helps to improve sharing of process information

supports the storage of BPM documentation in order to provide a single source of truth. Consequently, despite the internal changes, the BPM team has been able to maintain its focus and continue to develop BPM capability to optimize outcomes to the organization. Fig. 4 shows the major events, projects and capability development that have occurred within Company Q over the duration of its BPM Initiative.

4.3 Lessons Learned During the BPM Journey

Company Q has learnt many lessons during its BPM journey. These lessons relate to development and execution of strategies during the implementation of an organizational BPM approach. In keeping with the use of the BPM Capability Framework to underpin the BPM Initiative, Company Q maps the lessons learnt to the factors and capability areas to facilitate knowledge sharing and collaboration within the organization. The points in Table 3 provide an overview of the key lessons learnt by Company Q during its journey, mapped to the *factors* from the BPM Capability Framework.

5   Applying the BPM Capability Framework in Theory

In itself, the BPM Capability Framework is not a complete theoretical measurement model for BPM Initiatives as it does not include details of the relationships between and within the factors and capability areas nor does it include specific measurement items for the capability areas. To this end, the Principal Researcher is using the mappings and insights from the development of the Framework and its subsequent application, to further develop a theoretical measurement model as a part on the on-going program of research.

In 2007, the Principal Researcher conducted a second case study with Company Q. This case study was a part of a larger, longitudinal study undertaken with multiple organizations, with the purpose of investigating the progression of BPM initiatives, over time. The focus of this study was on (1) the emphasis that organizations, such as Company Q, had placed on the various capability areas including how this emphasis had changed over time and (2) the potential relationship between the factors and the capability areas with regard to the placement of emphasis. The Principal Researcher used the BPM Capability Framework to guide this investigation.

The appropriateness of the extension of the initial conceptual model, arising from the Delphi studies, becomes evident during the second case study with Company Q. In this study, the Principal Researcher calculated the change in emphasis placed on the capability areas, between two points, i.e. 2002 and then again for 2007[8]. This measurement used a 7-point scale with 1 being *Little or No Emphasis* and 7 being *High Emphasis*. The Principal Researcher then calculated the relative difference between the two scores for each capability area. Figure 5 and Figure 6 show the relative change in emphasis arising in both the factors and capability areas within Company Q for 2002 to 2007.

---

[8] This information came from data provided by a small number of Company Q employees at a single point in time (i.e. 2007). This approach to collecting longitudinal data presents potential limitations relating to the historic recollection of events for 2002 and the sample size. Within the larger study, the Researcher has addressed and mitigated these limitations. However, a full discussion of these points is outside the scope of this paper.

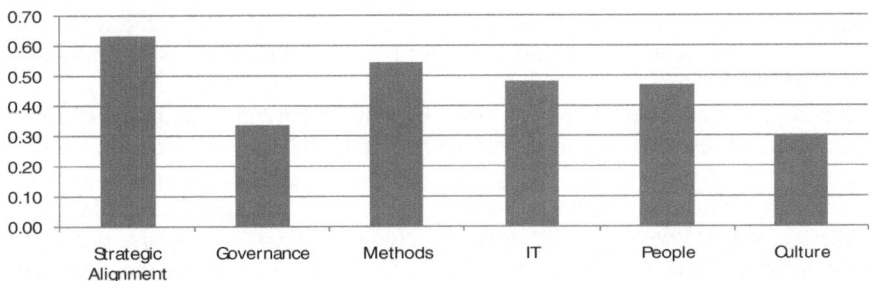

Fig. 5. Change in the Emphasis Company Q has placed on Factors from 2002 – 2007

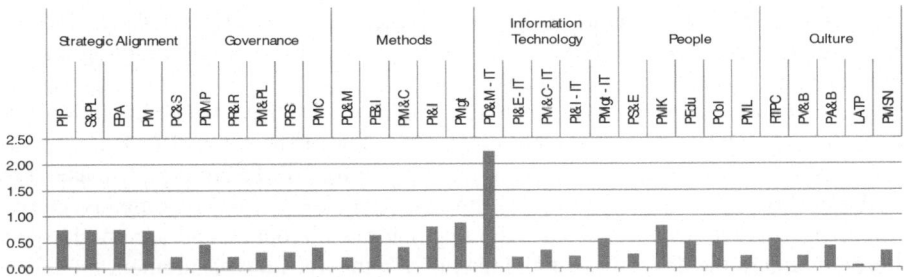

Fig. 6. Change in Emphasis Company Q has placed on Capability Areas from 2002 – 2007

Figure 5 shows that, from 2002 to 2007 the change in emphasis at a factor level ranges from 0.30 to 0.60. However, Figure 6 shows that for the same period the change in emphasis at a capability area level, ranges from 0.10 to 2.0, showing a much greater variation at the level of capability areas. From these figures, it is clear that consideration of the capability areas has the potential to provide greater insights than if only the factor level is considered. Furthermore, this data shows that within the capability areas, an organization places more or less emphasis on different capability areas over time. From a theoretical perspective, using the BPM Capability Framework enables researchers to identify potential relationships between, and within, both the factors and the capability areas. Furthermore, researchers are able to explore reasons for, and changes in, the relationships over time.

The research and experiences with Company Q raise interesting theoretical questions such as: What type of measurement model is most appropriate for measuring BPM Progression, reflective or formative? What are the temporal implications of progressing BPM capability areas? Is there a temporal order to BPM capability development? Are some capability areas better developed during early stages of an initiative and some later? Is this consistent across organizations, or can organizations get grouped based on commonalities in context (cf. [24])?

6 Conclusion

This paper reports practical and theoretical implications arising from the experiences of an Australian Transport Provider in adopting an organizational BPM approach. In particular, the paper presents the journey of Company Q's BPM Initiative and its participation in a research study undertaken to advance the progression and measurement of BPM Initiatives.

From a practical perspective, Company Q gained many insights into BPM by participating in the research and using the BPM Capability Framework to guide the development of its BPM journey. The practical experiences with Company Q show that the BPM Capability Framework is useful in providing structure and guidance to the progression of BPM within an organization. However, the lessons learnt during Company Q's BPM journey, indicate that often it is necessary to match the strategies for developing these capabilities to the individuals within, and the experiences of, the organization itself for them to be successful. The experiences of Company Q have shown the value in adopting a generic BPM Capability Framework to guide the progression of BPM. At the same time however, these experiences provide evidence of why a single, generic methodology is likely to be inappropriate for adopting and implementing an organizational BPM approach.

From a theoretical perspective, the Researchers at QUT have been able to integrate insights from industry into the ongoing development of theory on BPM progression including the development of a measurement model for BPM Maturity. The use of the BPM Capability Framework to assess the change in BPM within Company Q, over time, highlights the need to consider the implications of context and temporal aspects of progression in the development and testing of a theoretical measurement model.

By investigating and understanding such practical issues within future studies, researchers will be better positioned to contribute to the progression and sustainability of BPM within organizations. Furthermore, such insights increase the potential to develop more relevant, rigorous measurement models and theory suited to BPM as an organizational approach.

Acknowledgments

The authors wish to acknowledge the vital role of all Contributors to the on-going program of research that supports this paper. This includes the associated researchers from QUT and individuals from Company Q, without whom the insights presented in this paper would not have been possible.

References

1. Gartner Group: Meeting the Challenge: The 2009 CIO Agenda. EXP Premier Report Nr. Gartner Inc. Stamford, Connecticut (January 2009)
2. Pritchard, J.-P., Armistead, C.: Business process management - lessons from European business. BPMJ 5(1), 10–32 (1999)
3. Elzinga, D.J., Horak, T., Lee, C.-Y., Bruner, C.: Business Process Management: Survey and Methodology. IEEE Tran. Eng. Manag. 42(2), 119–128 (1995)
4. Armistead, C.: Principles of business process management. MSQ 6(4), 48–52 (1995)
5. Gulledge Jr., T.R., Sommer, R.A.: Business process management: public sector implications. BPMJ 8(4), 364–376 (2002)

6. Zairi, M.: Business process management: a boundaryless approach to modern competitiveness. BPMJ 3(1), 64–80 (1997)
7. Hammer, M.: The Process Enterprise: An Executive Perspective, http://www.hammerandco.com/publications.asp
8. DeToro, I., McCabe, T.: How to Stay Flexible and Elude Fads. QP 30(3), 55–60 (1997)
9. Llewellyn, N., Armistead, C.: Business process management: Exploring social capital within processes. IJSIM 11(3), 225–243 (2000)
10. Lee, R.G., Dale, B.G.: Business process management: a review and evaluation. BPMJ 4(3), 214–225 (1998)
11. Hatten, K.J., Rosenthal, S.R.: Managing the process-centred Enterprise. LRP 32(3), 293–310 (1999)
12. McDaniel, T.: Ten Pillars of Business Process Management. eAI Journal, 30–34 (November 2001)
13. Smith, H., Fingar, P.: Business Process Management – the Third Wave. Meghan-Kiffer Press, Tampa (2003)
14. Dumas, M., van der Aalst, W.M.P., ter Hofstede, A.H.M. (eds.): Process Aware Information Systems: Bridging People and Software Through Process Technology. John Wiley & Sons, New Jersey (2005)
15. Armistead, C., Machin, S.: Implications of business process management for operations management. IJOPM 17(9), 886–898 (1997)
16. Garvin, D.A.: The process of organisation and management. Sloan Manage. Rev. 39(4), 33–50 (1998)
17. Sabherwal, R., Hirschheim, R., Goles, T.: The Dynamics of Alignment: Insights from a Punctuated Equilibrium Model. Organ. Sci. 12(2), 179–197 (2001)
18. Harmon, P.: Business Process Architecture and the Process-Centric Company, http://www.buisnessprocesstrends.com
19. Rosemann, M., de Bruin, T., Hueffner, T.: A Model for Business Process Management Maturity. In: 15th Australasian Conference on Information Systems, Hobart, December 1-3 (2004)
20. Rosemann, M., de Bruin, T.: Towards a Business Process Management Maturity Model. In: 13th European Conference on Information Systems, Regensburg, Germany, May 26-28 (2005)
21. Rosemann, M., de Bruin, T.: Application of a Holistic Model for Determining BPM. In: AIM Pre-ICIS Workshop on Process Management and Information Systems, Washington D.C., pp. 46–60 (December 2004)
22. de Bruin, T., Rosemann, M.: Identifying BPM Capability Areas Using the Delphi Technique. In: 18th Australasian Conference on Information Systems Toowoomba, Australia, December 4-6 (2007)
23. Rosemann, M., de Bruin, T., Power, B.: A Model to Measure BPM Maturity and Improve Performance. In: Jeston, J., Nelis, J. (eds.) Business Process Management, ch. 27. Butterworth-Heinemann, Butterworth (2006)
24. de Bruin, T.: Insights into the Evolution of BPM in Organisations. In: 18th Australasian Conference on Information Systems, Toowoomba, Australia, December 4-6 (2007)

Collaborative Enterprise Modeling

Joseph Barjis, Gwendolyn L. Kolfschoten, and Alexander Verbraeck

Jaffalaan 5, 2628 BX Delft, The Netherlands
{J.Barjis,G.L.Kolfschoten,A.Verbraeck}@TUDelft.NL

Abstract. Key challenges in enterprise business process modeling are to capture complex inter-departmental and organizational processes, and to integrate different perspectives on the operation of the enterprise. Actors often convey different and only partly overlapping perceptions of their business processes, which hinder the construction of fairly accurate models in first modeling attempts. These different accounts of the business processes need to be integrated in a way to create a realistic and acceptable picture of the enterprise. To avoid this reoccurring pitfall and trial-and-error situation, and supporting the integration of different views on enterprise processes, collaborative modeling is emerging as a powerful approach. In this chapter, we report findings from a case study in which we used a collaborative approach to support enterprise business processes modeling with participation of analysts, process owners, and professionals. The deliverables of this chapter are based on a case study with participation of industry partners during a collaborative enterprise modeling session. We will reflect on the approaches used, lessons learned and the role of technology for supporting collaborative modeling.

Keywords: enterprise modeling, business process modeling, collaborative modeling, DEMO Methodology.

1 Introduction

A critical step in enterprise business process engineering and business process change is the analysis of the current business processes. To gain insight in the performance of the business processes and the challenges and opportunities for improvement, it is important that an integrated perspective of the current business processes is constructed. This integrated perspective requires not just input from all key stakeholders in the business process, but especially integration of these perspectives to resolve conflicts in understanding of the business processes. To support this integration, we present a study, discussion, and case on collaborative enterprise modeling using an interactive electronic whiteboard.

Enterprise modeling and business process analysis require to capture complex inter-departmental and organizational business processes, and to integrate different visions and perspectives on the operation of the enterprise in an integrated model (Prakash, 2008). When interviewed, actors often convey different and only partly overlapping perceptions of roles and tasks in business processes and it can be difficult to construct adequate models in first or second modeling rounds. Consequently, it is a

E. Proper, F. Harmsen, and J.L.G. Dietz (Eds.): PRET 2009, LNBIP 28, pp. 50–62, 2009.

challenge to distinguish the 'process as it is' (real day-to-day business) from the 'process that should be' (prescribed process) and the 'process that could be' (possible changes and innovation). Combined with the different views by different actors, it is highly complex to capture different perspectives on enterprise processes in an integrative model (Strazdina & Kirikova, 2008). To support integration of different views on enterprise processes, collaborative modeling has been proven to be a useful approach (Stirna & Kirikova, 2008). However, collaborative modeling itself is challenging, as it requires people within the business to express their views in terms of a modeling language. Therefore, collaborative modeling requires scientifically developed approaches, innovative collaborative technology, facilitation expertise, and appropriate modeling notations (Dennis et al, 1999; Persson, 2001; Stirna et al., 2007; Renger et al., 2008).

For a complete enterprise modeling activity, the models have to capture three phenomena; the enterprise processes, enterprise business rules, and enterprise information, which should be integrated in the corresponding change initiatives (Prakash, 2008). Thus, it is very important that enterprise models are accurate and complete, as flawed models will result in inadequate and unsupported change initiatives. Besides completeness, the enterprise model is also a way to communicate about the enterprise and the potential improvements that can be made. To engage stakeholders from the organization in the various phases of business process change, the image of their organization needs to be recognized by both the analysts and the stakeholders in the organization. Collaborative modeling is expected to support in creating a shared image of the organization that will simplify consecutive steps.

The goal of this chapter is to identify processes and guidelines on collaborative modeling to support enterprise modeling.

In this chapter we will report the findings from a case study in which we use a collaborative approach to support enterprise modeling and business process analysis with participation of analysts, industry partners (process owners), and professionals. We will reflect on the approaches used to support collaborative modeling and on the role of technology in enhancing and structuring the collaborative modeling effort.

In the remainder of this chapter we will first discuss the background and literature on collaborative modeling. Next we describe the modeling approach we used – the DEMO. Then we present the methodology of this study, followed by the case study description and its result. We conclude the chapter with a set of lessons learned followed by conclusions.

2 Collaborative Modeling

Collaborative modeling can be defined as the joint creation of a graphical representation of a system or process (Renger et al., 2008). Methods and approaches for collaborative modeling and group model building have been mainly developed in combination with problem structuring methods (Eden & Ackermann, 2006, Ackermann & Eden, 2005) and system dynamics modeling (Vennix, 1996, Andersen and Richardson, 1997, Rouwette et al., 2000). While these approaches are used to support problem solving and change in organizations they are based on a system perspective rather than a process perspective. Another line of collaborative modeling research in which process models are developed is found around the Group Support Systems research group in Arizona. In particular, Morton et al. (2003b) and Dean et al. (1994) worked on an approach for collaborative modeling concentrating facilitation support for the development of IDEF0 process models (Dennis et al., 1994, Hengst, 2005, Dennis et al., 1999).

Richardson and Andersen (1995) described five essential roles that should be present in collaborative modeling: the facilitator, modeler / reflector, process coach, recorder and gatekeeper. In this approach the modeler and facilitator construct the model in dialogue with the group. The recorder and process coach assist the facilitator in technology support and in the dynamics of individuals and subgroups. Finally the gatekeeper is the medium between the facilitation/modeling team and the participants from the organization.

A key objective in collaborative modeling is sense making. In order to create overlap in knowledge, participants need not only share information about the model elements and relations, they also need to create shared meaning with respect to these elements and their relations. Sensemaking is described by Weick (1995, p. 409) as involving "the ongoing retrospective development of plausible images that rationalize what people are doing." Sensemaking usually requires some development of shared meaning of concepts, labels, and terms. It also includes the development of a common understanding of context and the perspective of different stakeholders with respect to the model. The dialogue between modeler and stakeholders is critical in this process, to translate the tacit integrated perspective on the businesses process to a modeling language. In this study we used DEMO business transactions as the modeling language.

3 The Modeling Language

For the modeling purpose, we adapted the DEMO (Design and Engineering Methodology for Organizations) modeling language that has been used in enterprise modeling (Dietz, 2006). The DEMO modeling language is chosen because it derives abstract models representing only essential activities and is based on formal semantics. Below we briefly introduce the DEMO Methodology ontological transaction concept and the diagrammatic representation of a business transaction based on this concept. The interested reader is referred to the book by the author of the DEMO Methodology (Dietz, 2006). This methodology takes a philosophical stance that an enterprise is first of all a social system where human actors interact in order to fulfill the organization mission.

Social actors in organizations perform two kinds of acts: *production act* (P-acts, for short) and *coordination acts* (C-acts, for short). By engaging in P-acts, the actors bring about new results or facts, e.g., they deliver a service or produce goods. By engaging in C-acts, the actors enter into communication, negotiation, or commitment towards each other. The generic pattern in which the two kinds of actions (P-acts and C-acts) occur is called a *transaction*, see Figure 1. In the terminologies used, we return to earlier terms used for P-acts and C-acts, i.e., instead of P-act, we refer to it as *action*, and C-act as *interaction*.

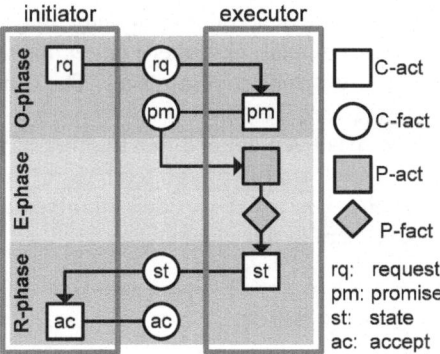

Fig. 1. Basic Transaction Concept (adapted from Dietz 2006)

A business transaction is a pattern of action and interaction. An *action* is the core of a business transaction and represents an activity that brings about a new result. An *interaction* is a communicative act involving two actor roles to coordinate or negotiate a particular action. Examples of an interaction could be:

- "requesting a new insurance policy"
- "clicking an apply/submit button on an electronic form"
- "inserting bank card into an ATM to withdraw cash"
- "pushing an elevator's summon button".

Replying to the interacting actors and fulfilling their requests is an action, for example:

- "issuing a new policy"
- "processing an e-form"
- "dispensing bills"
- "moving an elevator to the corresponding floor".

Another notion of the transaction concept is the actors role involved in a transaction. Each business transaction is carried out by exactly two actor roles (initiator and executor), as illustrated in Figure 1. As seen from the figure, a business transaction is carried out in three distinct phases:

- *Order phase (O)*, during which an actor makes a 'request' for a service or good towards another actor. This phase may include a number of communicative acts or 'interactions' (negotiation, discussion). This phase ends with a commitment ('promise') made by the second actor, who will eventually deliver the requested service or good. This phase represents 'interaction'.
- *Execution phase (E)*, during which the second actor fulfills its commitment, i.e., 'produce' the service or good. This phase includes a productive act or 'action', carried out solely by one actor, the executor.
- *Result phase (R)*, during which the second actor does 'state' to the first actor that the service or good is produced. This phase also may include a number of communicative acts or 'interactions' (negotiation, discussion). This phase ends with the 'accept' of the service or good by the first actor. This phase also represents 'interaction'.

It should be noted that at any point (phase) an actor may quit the process (e.g., canceling the request, refusing the result presented) or decline to proceed, or a process is terminated due to internal or external circumstances. The cancellation pattern is not depicted, but can be found in (Dietz, 2006).

The DEMO Methodology has been applied to a number of life systems, including complex enterprise systems (see for a list of publications on the website of the methodology: http://www.demo.nl/). One of the studies involves a small-medium enterprise, which will be discussed here in this chapter.

4   Method and Case Study Setup

An in-depth case study was conducted on Kemeling Kunststoffen BV, a plastic production company, located in Westland-Area, The Netherlands. The company recently launched an initiative to review its business processes and improve the current processes by reducing delays, and developing new information systems that will support the redesigned business processes. In particular, the enterprise wanted a business process monitoring system.

For this case study we used an action research approach. Action research is used more often in collaborative modeling research as it requires modelers (researchers) to make an active intervention in a group of participants, to study the effect (Morton et al., 2003a). As modeling requires highly advanced skills, it is difficult to train others to make this intervention to observe its effect.

We used the action research cycle from Baskerville and Wood-Harper (1996) (1) diagnosing (2) action planning, (3) action taking, (4) evaluating and (5) specifying learning. In the diagnosing stage, the key problems that require the enterprise to change or improve are identified. In the action planning step, the intervention is designed. The researchers and industry partners (business process owners) envision an approach to change the situation. In the action taking step, the actual intervention is made. In the evaluation we reflect whether the change had the desired, and theoretically predicted effect. Finally the industry partners and researchers reflect on what they learned from the intervention.

In this study we wanted to compare a traditional enterprise modeling effort with a collaborative approach. The traditional modeling was done by students. They visited the organization and performed semi-structured interviews to create a formal demo model of the enterprise. The model was captured in a report with a description and handed to the company stakeholders. Next the collaborative modeling effort was organized as described below.

4.1   Participants of the Collaborative Session

The collaborative enterprise modeling approach we applied for conducting enterprise modeling of Kemeling Kunststoffen BV was carried out as follows.
We organized a half day session with the following participants:

- Kemeling Kunststoffen BV– Two senior employees (managers) with extensive knowledge of the operations and daily routines in the enterprise.
- Enterprise modeling expert – an author of enterprise modeling methodology

-     Modelers – two simulation and modeling experts
-     Facilitator – a professional facilitator.
-     Observers – a group of observers to document the session.
-     Graduate students – a group of graduate students who successfully completed an enterprise modeling course. These students prepared a preliminary business processes description, that served as a starting point for the collaborative session. The students study was completed as part of their graduate course work.

The role of recorder and gatekeeper and process coach were fulfilled by the modeler and facilitator to keep the modeling team of a reasonable size. All together, in the session we had 9 people.

4.2   The Action Research Cycle

Diagnosing
The company had identified several processes where product delivery can be streamlined and the production cycle can be shortened. Also the company was hoping to see the improvements realized through business process redesign and IT support. As the company is growing, the system becomes more complex and there is a need to better understand the enterprise business processes. Further, as a basis for change, it is important that the stakeholders in the company and the analysts have a shared understanding of the enterprise.

Action Planning
One of the researchers involved took the role of modeling expert, and one of the researchers took the role of facilitator. First we explained the industry partners the purpose of the modeling exercise. Next we explained the modeling approach and created a small model to illustrate the approach with an example/tutorial to explain the key elements and relations used to build the model. From the modeling language we only used the transactions, actors and results to keep the model as simple as possible. We kept the modeling notations to 2-3 elements (box, arrow, swim-lane), but enhanced them with color to visually distinguish between actions and results modeled. To support the collaborative modeling effort we used an electronic whiteboard with smart ideas™ software. This tool allowed us to capture the model during the collaboration process in a way that is readable for all participants and flexible with respect to corrections and layout. Given the size of the model a normal whiteboard or flip-over would have been full and difficult to read. Further, when making changes or corrections the model would become messy. Using the interactive whiteboard the model could be modified and saved digitally. The set-up used for the modeling task is visualized in Figure 2.

In this set-up both the smartboard and the projector were coupled to the workstation. Additional, a wireless keyboard was attached. The facilitator could manipulate the model using the interactive whiteboard (touch screen) or the wireless keyboard. All participants could see the full size model at the projection screen.

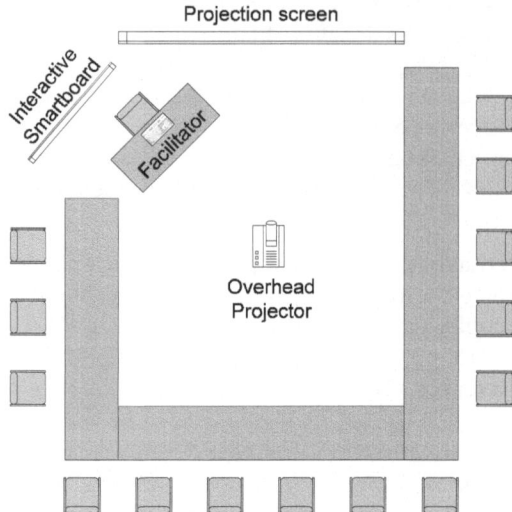

Fig. 2. Set-up for the Collaborative modeling effort

Action Taking

Based on a description of their business processes one of the modeling experts identi-fied the main business transactions in the organization. The main business transaction was to order and deliver products, and to pay for products. From this point, more detail was created, adding actors and transactions such as designing the product, pro-duction, ordering supplies and handling defective products. Each transaction was captured directly in the model, and each time it was verified that the customers under-stood the model, and the meaning of the model. When the level of detail increased, the discussion with the customers also increased, verifying how transactions took place, and asking for clarifications and examples. When the expert felt the model was reaching completeness, the clients were asked if they were missing aspects of the identified problems. Modeling experts involved in the analysis were asked to add transactions they encountered. These transactions were recorded, and it was discussed whether they were actually new transactions or covered as part of other transactions. This process continued for about 80 minutes, which resulted in a complete model, but still a little bit messy and disorganized (see Figure 3). But the model was simple enough for everyone to read and understand, and it was agreed upon by all partici-pants (analysts, experts, industry partners) that the model was, for now, correct. Due to the size of the diagram represented in Figure 3, the texts in the boxes may not be readable. The intention is to show how a free format was used to quickly capture the business transactions and business processes. It is not meant to be readable to the details. The figure shows how the business transactions are interrelated in a network and what are the initiator and executor roles for the transactions. On the left-bottom of the figure, we listed further business transactions (T17-T21) that are also relevant to the process such as possible customer complaints and their handling, resolving pay-ment dispute, etc.

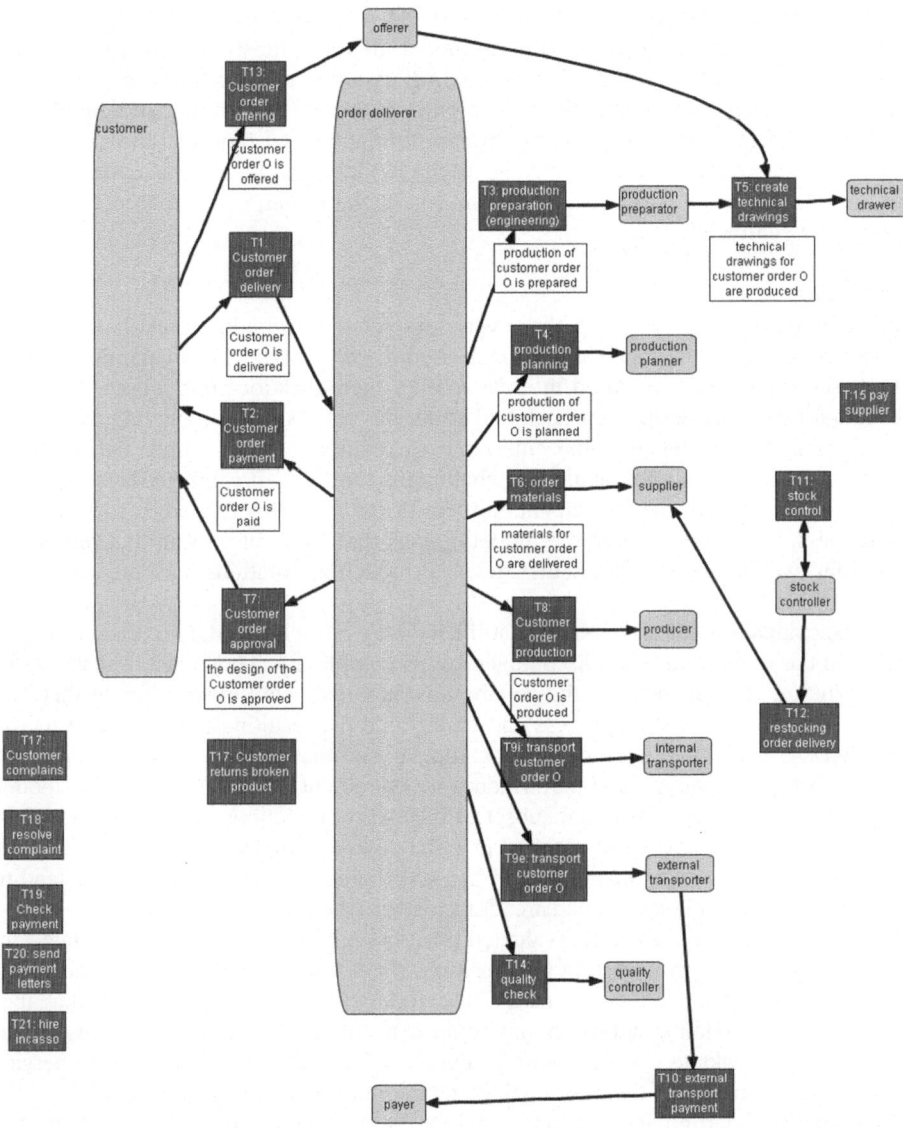

Fig. 3. Kemeling Kunststoffen BV enterprise model constructed collaboratively

As we mentioned above, we wanted to keep the modeling notations really simple so everyone could follow it. At the same time, in addition to modeling, we were taking notes whether a transaction is simple, composite, optional, etc. These notes and the model of Figure 3 helped to produce a detailed model (not included in this chapter) using the more extensive notations. These extended notation are not o relevant to this chapter, however the interested reader is referred to (Barjis, 2007), where they can find a case study with the use of extended notations and simulation.

During the collaborative modeling effort the industry partners and the modeling experts did not directly interact with the model, this was mostly done by the facilitator, and occasionally by one of the modeling experts. The facilitator captured new transactions when mentioned. After discussion between the industry participants and modeling experts, they agreed on the transaction labels and representation, and these were edited to create a shared model. After the model construction was completed, we had a "discussion and feedback" session where the collaborative process was discussed and evaluated.

Evaluation
Immediately after the modeling phase was completed, we conducted evaluation of the whole modeling session and the approach used to collaborate. The evaluation method was based on a semi-structured interview. We asked questions both about the report of the traditional modeling approach and about the collaborative modeling effort. The questions addressed whether the enterprise stakeholders could follow the modeling process, and if no, elaborate what was challenging and how it can be addressed. Also, we discussed whether the participants felt comfortable and motivated to participate in the process, and the aspects of the modeling task that were relevant in that respect. In particular, we wanted to see whether the used modeling notations presented any challenge for them.

In the evaluation, it became clear that the industry participants did not understand the model in the report and were discouraged by its complexity. They made remarks comparing the reported model and the collaboratively constructed model, saying that they now followed the modeling process and understood the resulting model and its merits. The client indicated that the model was complete, and that they appreciated its value as a basis for further analysis and performance measurement. While the traditional modeling approach resulted in different subset of business transactions in every round of the model, the collaborative effort resulted in a shared and complete model of the relevant business transaction. The annalists and the stakeholders gained shared understanding about the abstraction level and demarcations made in the modeling effort.

The process and set-up were evaluated positively. The stakeholders appreciated the introductory tutorial on modeling that prepared them to feel comfortable and confident to participate in the modeling process. The industry participants indicated that they understood the modeling approach and could read the model without problems, which was achieved thanks to simple notations and use of interactive whiteboard. Witnessing the actual construction of the model helped them to understand its meaning.

The industry participants indicated that the (static) model did not teach them new things as they couldn't see how the current business processes or actor-interaction can be altered and manipulated. They mentioned that a simulation of the model would help them to see the processes in a dynamic way and would enable them to manipulate some of the parameters.

Learning
Below we mention the key lessons learned from the joint session with industry.

- The modeling language used should be as simple as possible to capture the complexity required. Although Kemeling Kunststoffen BV is an SME (small-medium enterprise), it required significant effort to capture all relevant business transaction. We found that while the on-site interviews offered a basic understanding of the enterprise, the resulting model was too complex for the organizational stakeholders to recognize their business transactions. Traditional analysis at an enterprise with distributed structure and more versatile and agile environment will pose much greater challenges to be dealt with if enterprise modeling is conducted merely in an interview-like manner without collaborative sessions. To involve the stakeholders the modeling language need to be simple.

- The participating stakeholders indicated that the modeling effort helped them to understand the resulting model. Seeing a model being built is easier to understand than trying to understand a completed model presented by an expert. Based on the resulting model the analysts could create a more complex notation, while keeping a clear and recognizable picture that captured shared understanding between the enterprise stakeholders and the analysts.
- The modeling language should be explained using an example. The stakeholders from industry are not familiar with the particular modeling language, and explaining it during the actual modeling effort could result in misunderstanding. To avoid this, it is useful to create a small example model, of 3-5 elements and relations. In this way the modeling language can be explained.
- It is important to explain the need for modeling and its role in the process of business improvement. Being involved in the modeling effort requires the stakeholders to learn the modeling language and to engage in the modeling effort. It is important for the stakeholders to understand why this is important. Participants from industry intuitively focus on the solution, brainstorm, rather than the analysis of the current process and its bottlenecks. To get them engaged in the modeling effort they have to understand the need for analysis and the value of shared understanding on the current enterprise system and processes for later improvements.
- Completeness and correctness of the model is more likely when the business owners (managers) are involved in the actual modeling effort. In confronting different perspectives of the organization and the analysts, tradeoffs and different views are integrated or resolved. Further, when the model is created with all relevant stakeholders, completeness is more likely.

5 Discussion

From the case study we learned that the collaborative approach to conceptual modelling and analysis of the enterprise improves the completeness, correctness and shared understanding of the system and related processes. In this section we will argue that this investment is likely to improve the quality and efficiency of all phases in a business change process, especially when this process involves complex simulation and experimenting to identify problems and to experiment with the effects of improvements.

Business process change uses the process re-engineering life cycle to describe the process from envisioning to inauguration, to diagnosis, to (re-) design, to (re) construction and to evaluation (Grover and Kettinger 1995; Kettinger and Teng 1997).

The inauguration phase is the phase in which the investment decision is made. Is it worth while to make changes in the enterprise? Once it is determined that change is required, it is important to diagnose the problems and bottlenecks in the organization. For this diagnoses a detailed analysis of the current system is required. This phase consists therefore of a conceptual modeling phase, followed by the creation of a simulation model to investigate how resources are used over time, and where in the enterprise process or system delays and bottlenecks can be found. Once these are found, the stakeholders and analysis have to come up with solutions and design changes to the system. Literature is very explicit about the need to involve both designers and users or stakeholders when designing change (Standish Group 1995; Standish Group 1996; Boehm, Grünbacher et al. 2001; Acosta and Guerrero 2006). Once all stakeholders support the envisioned change, it can be constructed, implemented in the organization, after which it is evaluated to make further improvements and identify new problems.

In the conceptual phase the collaborative modeling effort can help to create a shared understanding between the analysis and design team and the users and stakeholders of the enterprise. Using a shared image of the processes and systems in the organization will improve the quality and efficiency of consecutive steps. For instance once there is shared understanding on the conceptual model, the stakeholders of the organization are more likely to recognize their business processes and systems in the simulation model that is created. Labels of various concepts in the enterprise are more consistent, relations are more accurate, and the model has agreed-on demarcations and levels of abstraction. Based on this simulation model it is easier to discuss bottlenecks and challenges, as they link more directly to real elements and concepts in the enterprise. Once challenges are found, the achieved shared understanding will also improve the finding of solutions. As problems are linked to real elements in the enterprise, it becomes easier to involve the right stakeholders in finding solutions. This involvement in finding and designing solutions is again critical to gain support for the changes in the organization. Resistance to change can be overcome when users and stakeholders are involved in the design phase. This will in turn make implementation smoother. In the evaluation phase again, stakeholders and analysts will benefit from their shared understanding to evaluate change and it's effects.

6 Conclusions and Further Research

In this chapter we presented a collaborative approach to enterprise modeling based on a case study in which we used action research to learn from our interventions. From our case study we learned that collaborative modeling improves the shared understanding among stakeholders of the enterprise and analysts. We found that it takes some investment of business stakeholders to engage in the collaborative modeling effort, but we argue that this is likely to be worthwhile as consecutive steps are likely to benefit from the increased shared understanding. Based on what we learned from this preliminary case study, we propose to structure the collaborative modeling process in three phases: *tutorial* (introducing participants to the modeling notations and concepts to be used), *modeling* (modeling of the enterprise business processes), *change* (analysis and improvement of the existing practice). Each of these three phases requires detailed guidelines and step-wise facilitation scenarios to be developed.

The study presented in this chapter has two key limitations. First and most important the limited size of the case study, in which only one case and a small enterprise was studied. Second, the prolonging effect of the collaborative modeling effort was not yet evaluated.

A future research should aim to further develop and test this collaborative approach with enhancements based on the lessons we have learned and apply the approach to more complex enterprises. We will see if such a modeling session will lead to better results, satisfaction and whether it addresses the needs of the industry partners.

Second, we would like to formally describe the approach as a set of design patterns for collaborative modeling. Collaborative work practices can be described with design patterns. An example of this can be found in the use of ThinkLets (Vreede et al. 2006).

Third we would like to study and measure the effect of the joint model on consecutive phases in the business process reengineering process. We argue in our discussion that it should improve the quality and efficiency of future steps, and should increase the support for the solutions and changes to the enterprise.

References

Acosta, C.E., Guerrero, L.A.: Supporting the Collaborative Collection of User's Requirements. In: International Conference on Group Decision and Negotiation. Universitätsverlag Karlsruhe, Karlsruhe (2006)

Ackermann, F., Eden, C.: Using Causal Mapping with Group Support Systems to Elicit an Understanding of Failure in Complex Projects: Some Implications for Organizational Research. Group Decision and Negotiation 14, 355–376 (2005)

Andersen, D.F., Richardson, G.P.: Scripts for Group Model Building. System Dynamics Review 13, 107–129 (1997)

Barjis, J.: Automatic Business Process Analysis and Simulation Based on DEMO. Journal of Enterprise Information Systems 1(4), 365–381 (2007)

Baskerville, R.L., Wood-Harper, A.T.: A Critical Perspective on Action Research as a Method for Information Systems Research. Journal of Information Technology 11, 235–246 (1996)

Boehm, B., Grünbacher, P., Briggs, R.O.: Developing Groupware for Requirements Negotiation: Lessons Learned. IEEE Software 18(3) (2001)

Dean, D.L., Lee, J.D., Orwig, R.E., Vogel, D.R.: Technological Support for Group Process Modeling. Journal of Management Information Systems 11(3), 43–63 (1994)

Dennis, A.R., Hayes, G.S., Daniels Jr., R.M.: Re-engineering Business Process Modeling. In: Proceedings of the Twenty-Seventh Annual Hawaii International Conference on System Sciences (1994)

Dennis, A.R., Hayes, G.S., Daniels Jr., R.M.: Business process modeling with group support systems. Journal of Management Information Systems 15(4), 115–142 (1999)

Dietz, J.L.G.: Enterprise Ontology –Theory and Methodology. Springer, Heidelberg (2006)

Eden, C., Ackermann, F.: Where Next for Problem Structuring Methods. Journal of the Operational Research Society 57, 766–768 (2006)

Grover, V., Kettinger, W.J.: Business Process Change; Reengineering Concepts, Methods and Technologies. Idea group Publishing, Harrisburg (1995)

den Hengst, M.: Collaborative Modeling of Processes: What Facilitation Support does a Group Need? In: Americas Conference on Information Systems. AIS Press, Omaha (2005)

Kettinger, W.J., Teng, J.T.C.: Buisiness process change: a study of methodologies, techniques and tools. Management Information Systems Quarterly 21(1), 55–80 (1997)

Morton, A., Ackermann, F., Belton, V.: Technolog-Driven and Model-Driven Approaches to Group Deccision Support: Focus, Research Philosophy, and Key Concepts. European Journal of Information Systems 12, 110–126 (2003a)

Morton, A., Ackermann, F., Belton, V.: Technology-driven and model-driven approaches to group decision support: focus, research philosophy, and key concepts. European Journal of Information Systems 12(2), 110–126 (2003b)

Persson, A.: Enterprise Modelling in Practice: Situational Factors and their Influence on Adopting a Participative Approach, PhD dissertation, Dept. of Computer and Systems Sciences, Stockholm University (2001)

Prakash, N.: Bringing Enterprise Business Processes into Information System Products. Bringing Enterpris. In: Stirna, J., Persson, A. (eds.) PoEM 2008. LNBIP, vol. 15, pp. 168–181. Springer, Heidelberg (2008)

Renger, D.R.M., Kolfschoten, G.L., de Vreede, G.J.: Patterns in Collaborative Modelling: A Literature Review. In: Climaco, J., Kersten, G.E., Costa, J.P. (eds.) Group Decision and Negotiation, Coimbra, Portugal, Faculdade de Economia da Universidade de Coimbra (2008)

Richardson, G.P., Andersen, D.F.: Teamwork in Group Model Building. System Dynamics Review 11(2), 113–137 (1995)

Rouwette, E.A.J.A., Vennix, J.A.M., Thijssen, C.M.: Group Model Building: A Decision Room Approach. Simulation & Gaming 31(3), 359–379 (2000)

Standish Group. CHAOS Report: Application Project and Failure (1995)

Standish Group, The International Unfinished Voyages, A Follow-Up to The CHAOS Report (1996)

Stirna, J., Kirikova, M.: Integrating Agile Modeling with Participative Enterprise Modeling. In: The proceedings of the CAiSE workshop EMMSAD (2008)

Stirna, J., Persson, A., Sandkuhl, K.: Participative Enterprise Modelling: Experiences and Recommendations. In: Krogstie, J., Opdahl, A.L., Sindre, G. (eds.) CAiSE 2007 and WES 2007. LNCS, vol. 4495, pp. 546–560. Springer, Heidelberg (2007)

Strazdina, R., Kirikova, M.: Business process modeling perspective analysis. In: Stirna, J., Persson, A. (eds.) PoEM 2008. LNBIP, vol. 15, pp. 210–216. Springer, Heidelberg (2008)

Vennix, J.A.M.: Group Model Building: Facilitating Team Learning Using System Dynamics. John Wiley & Sons, Chichester (1996)

Vreede, G.J., de Briggs, R.O., Kolfschoten, G.L.: ThinkLets: A Pattern Language for Facilitated and Practitioner-Guided Collaboration Processes. International Journal of Computer Applications in Technology 25(2/3), 140–154 (2006)

Weick, K.E.: Sensemaking in Organizations. Sage Publications Inc., Thousand Oaks (1995)

Assessing the Efficiency of the Enterprise Architecture Function

Bas van der Raadt[1,*] and Hans van Vliet[2]

[1] Capgemini, Global Financial Services / Architecture & Governance Improvement,
Papendorpseweg 100, 3528 BJ Utrecht, the Netherlands
bas@vanderraadt.nl
[2] VU University, Department of Computer Science
De Boelelaan 1081a, 1081 HV Amsterdam, the Netherlands
hans@cs.vu.nl

Abstract. Many large organizations have difficulties managing enterprise transformations involving their extensive and complex portfolio of business processes, information systems, and infrastructure. Enterprise Architecture (EA) is an increasingly important instrument to better manage enterprise transformations. EA provides a means for getting insight into the current state landscape, creating a target blueprint, and setting out a roadmap to achieve that target state. Although investing heavily in EA, few organizations have implemented a truly efficient EA function. In order to implement an improvement cycle for the EA function, organizations conduct efficiency assessments. In this article we present an integral assessment model to determine the efficiency of the entire EA function. Our model takes an eclectic view, which makes it easily adaptable to organization specific characteristics. We use a case description to illustrate the working of our model, and to show which type of concrete insights it provides to identify points for improvement.

Keywords: Enterprise Architecture, Function, Efficiency, Organizational Assessment.

1 Introduction

Enterprise Architecture (EA) is a multipurpose instrument, which is increasingly being used by large organizations to coordinate and manage enterprise-wide transformations of their complex landscape of business and IT assets distributed across a geographically diverse organizational structure [1], [2]. EA provides a means for creating a target blueprint [3] for the entire enterprise or a specific domain. It provides a logical representation – through models and descriptions – of the future business functions, activities, processes, information needs and systems, and technical infrastructure as a foundation for executing the business strategy [4]. Making an inventory of the current landscape and performing a gap-analysis against the target blueprint,

---

* As of April 1st 2009, Bas van der Raadt is employed at Ernst & Young Netherlands, Program Advisory Services.

E. Proper, F. Harmsen, and J.L.G. Dietz (Eds.): PRET 2009, LNBIP 28, pp. 63–83, 2009.

allows identifying the roadmap [5] for achieving the desired situation. The target blueprint acts as a common frame of reference for managing and coordinating an enterprise transformation. All projects and changes that are part of the enterprise transformation are validated on their conformance to the target blueprint, and deviations are transparently managed through waivers and escalations.

Many organizations have been applying EA for some time, and have one or more teams of enterprise architects working for them [6]. Typically, enterprise architects are quite experienced employees, often highly valued for their knowledge about structures, processes, systems and technology of the organization. This architectural knowledge supports various stakeholders in their decision-making [7] and implementation processes. However, the EA functions of many organizations are not fully efficient yet. This often results into these organizations not being effective in using EA as an enterprise transformation management instrument. There are many factors that determine the efficiency or inefficiency of the EA function. For example, the architects create too abstract and too high-level enterprise architectures, which provide little concrete information to coordinate the projects as part of an enterprise transformation. Architectures created at project level, on the other hand, are often too detailed. They do not provide the required overview of, and insight in the interrelations between, the individual projects as part of a total enterprise transformation. Another common problem is that the organization's (or transformation program's) governance structure and processes are immature, making it hard for EA decision making to be enforced in such a way that EA conformance of projects and changes is ensured [8]. The result is that EA products are often hardly used for what they were intended for; they often end up as shelf-ware and are hardly put into practice [9].

The literature provides various models for performing organizational assessments describing factors that determine EA efficiency. Typically, these models focus on the EA delivery function – e.g., [6], [10]. The EA delivery function is the team of architects responsible for creating and maintaining EA products (architectures and EA policies). In our view, the EA function reaches beyond the EA delivery function, and also includes the bodies, roles, structures and processes involved with ratifying, enforcing and conforming to the EA products [8]. The scope of existing EA assessment models does not typically include these elements.

Most existing EA assessment models focus on determining process maturity, and are therefore process oriented – e.g., [11], [12], [13]. These models describe the characteristics of several maturity phases an EA function passes while becoming more efficient. They describe a typical pattern of EA efficiency development, assuming that a certain maturity phase, regardless of the unique characteristics of an organization, typically involves a specific description of various efficiency topics (e.g., process standardization, linkage to business strategy, management involvement, etc.). However, the development path of EA within organizations may differ for each organization [14], which means in specific situations these patterns may not apply. Also, incorporating various topics into one maturity phase makes it hard, or even impossible, to assess all topics individually in order to identify improvement points.

In this article, we present a model, as part of our Normalized Architecture Organization Maturity Index (NAOMI), for assessing the efficiency of the full scope of the EA function. This model ensures that an assessment provides complete insight in the efficiency of all organizational functions, roles, and bodies involved in creating, maintaining, ratifying, enforcing, and conforming to Enterprise Architecture decision-making. The model does not link its efficiency topics to specific maturity phases. Our approach consists of an EA function reference model, providing a norm description of the EA decision making, delivery, and conformance activities of the EA function [8]. Therefore, our model is flexible and can be applied in various situations, providing standard efficiency profiles. These profiles and other outcomes of our assessment approach provide useful insights, based on which points for improvement can be identified to maximize EA function efficiency.

This article is structured as follows. Section 2 describes our research approach. Section 3 focuses on the efficiency of the EA function. Section 3.1 elaborates on the efficiency construct. In Section 3.2 we briefly discuss our reference model which provides a norm description of an EA function. Section 4 provides a detailed description of our assessment model describing the efficiency topics for the entire EA function (Section 4.1) and specifically for the EA delivery function (Section 4.2). In Section 5 we briefly discuss our standard assessment approach. In Section 6 we illustrate our EA function assessment model and approach using a case description of an assessment we conducted. In Section 7 we discuss the lessons learned from these assessments and related work. We conclude, in Section 8, with the key findings and recommendations for future research.

2 Research Approach

We created our EA function efficiency model using the following approach. We started with performing a literature review of existing assessment models and approaches (i.e. [6], [10], [11], [12], [13]). We analyzed their strong points as well as, in our view, their points of improvement – e.g., gaps in the assessment models, impracticalities in the approaches, etc. (see related work in Section 7.3). We created an assessment model combining the strong points of the existing models. We filled the gaps we identified with new elements with literature from various research fields, resulting in the first version of our assessment model [15].

We conducted two case studies at financial services companies (i.e. [16], [17]) to qualitatively test the validity of this first version of our assessment model. From conducting these case studies we learned that in order to conduct a complete assessment of the EA function, some key elements were still missing. Based on these lessons learned we improved our assessment model and approach.

First of all, we found that existing EA assessment approaches typically focus on determining the efficiency of the EA delivery function – i.e. the team or department of architects responsible for creating the architectures and EA policies. We learned that other functions and roles within organizations also have tasks and responsibilities in the EA function. For example, senior management has an important responsibility regarding EA decision making. Project members such as project managers, analysts and designers have the responsibility of working according to the EA products. Therefore we increased the scope of our EA function assessment model to also include these elements of EA decision making and EA conformance.

Secondly, we found that, as a result of their too narrow scope, existing assessment models do not clearly describe required governance, collaboration and coordination structures and processes for the entire EA function. Also, they do not explicitly take into account the key parameters of an EA function to determine how it should function in order to be efficient and effective – e.g., the positioning of the EA function in the organization chart, and its coverage of architectural domains. Therefore, we included these key parameters for the entire EA function, including its positioning, governance, and collaboration and coordination into our assessment model (see Section 4.1).

Thirdly, existing EA assessment approaches do not explicitly describe a separate reference model for the EA function. They make no clear distinction between the EA function assessment measure (the topics to be assessed) and the EA function reference model (the norm to compare a specific EA function with). Therefore, we separated the reference model (see Section 3.2) that describes a generic model of the EA function from the assessment model (see Section 4) that may be used to determine the gap between a specific EA function and the generic reference model.

These changes resulted in a new version of our assessment model, which we describe in detail in Section 4 of this article. We conducted another case study to qualitatively test this version of the model (see Section 6).

3   Efficiency of the EA Function

3.1   Efficiency

In the literature, terms like effectiveness, efficiency, and effort are often not clearly defined and distinguished. In order to clearly define which concepts our model assesses, we shortly elaborate on these concepts. *EA effectiveness* is the measure of goal achievement [18] with EA – e.g., reducing the complexity of an organization, regarding both business and IT. Other examples are improving the agility [19] and business IT alignment [20] of the organization. *EA efficiency* concerns the quality of the architecture process in term of execution time and accuracy, for example in providing advice to management through EA. A way to increase efficiency of the EA function is to standardize its processes, train its architects, and improve its management and communication capabilities. Efficiency is a predictor of outcome effectiveness [21]. *EA effort* relates to the way the resources of the EA function are consumed [18], such as money and energy. Effort involves those activities that do not change the EA function and its processes, but merely increase the amount of resources (e.g. architects, tools, hardware, etc.) to decrease the execution time and number of errors.

3.2   EA Function Reference Model

In our reference model (described in detail in [8]), we define the EA function as: *"The organizational functions, roles and bodies involved with creating, maintaining, ratifying, enforcing, and observing Enterprise Architecture decision-making ...established in EA products ...interacting through formal (governance) and informal (collaboration) processes at enterprise, domain, project, and operational levels".*

EA products describe the architectural decisions taken, and provide a means for communicating and enforcing these decisions throughout the organization. There are generally two types of EA products: (1) architectures and (2) EA policies [22]. An *architecture* document provides an abstraction of what a complex environment looks like, and acts as a means of communication and decision making regarding that environment [14]. An *EA policy* prescribes how projects should implement organizational changes across various subunits through unified principles and practices [8], which allows organizations to centrally control the change activities of subunits without dictating exactly how they handle the details [2].

Our EA function reference model describes three core activities: (1) EA decision making, (2) EA delivery, and (3) EA conformance [8]. *EA decision making* is concerned with approving new EA products or changing existing EA products. Also, it deals with resolving conflicts between the functions, bodies and roles within the EA function, and is responsible for resolving issues of non-conformance to EA products. *EA delivery* is responsible for creating and maintaining EA products, and provides advice to senior management to guide EA decision making. EA delivery also validates projects and operational changes to see whether they conform to the EA, and provides support in applying EA products. *EA conformance* is responsible for running change projects and implementing operational changes as described in the target architectures, complying with the EA policies. EA conformance also includes providing feedback on the applicability of the EA products.

4 EA Function Efficiency Assessment Model

Our EA function efficiency assessment model is divided into two parts. Part 1 describes at the level of the entire EA function which are the essential elements of an efficient EA function (see Section 4.1). Part 2 describes the efficiency variables that are specific for the EA delivery function (see Section 4.2).

4.1 Part 1: The Entire EA Function

Based on two case studies [16], [17] performed, we found that the EA function needs to be clearly defined regarding its position, strategy, structure and operating model, and all stakeholders involved must be made aware of this definition. Also, the three core activities of the EA function should operate as one unity, and thus needs to be well governed.

Peterson [23] describes three essential capabilities for a well governed organizational function – structural, process, and relational capabilities – which fit our findings from the case studies we performed. We used Peterson's fundamental work in the field of IT governance to set up the governance, collaboration and coordination aspects of our EA function reference model [8]. We translated these capabilities into three essential preconditions for EA function efficiency: (1) a clear and accepted EA function definition, (2) a transparently and consistently operating EA governance

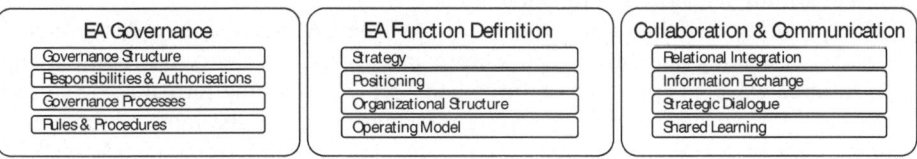

Fig. 1. Assessment topics for determining the efficiency of the EA function

model, and (3) proactive collaboration and communication between all functions, bodies, and roles that take part in the EA function. We used these three key preconditions to construct our EA function efficiency assessment model shown in Figure 1.

4.1.1 Definition of the EA Function

The EA function should have a well described *strategy*, which states its mission, purpose and objectives. The strategy should be based on the positioning of the EA function, and should be well-aligned with the corporate strategy. For example, for the back-office of a large international company we assessed (see Section 6), complexity of the operational information systems and infrastructure are the primary concern. The EA function's mission is to ensure that projects and operational changes add to the simplification of the operational environment. Typical objectives of an EA function are to reduce costs, complexity and risk, and increase flexibility and quality of service regarding change projects and the operational environment [4].

In order to realize its strategy, the EA function should have the right position within the organization. The EA function's *positioning* consists of three variables: (1) organizational scope, (2) organizational levels, and (3) architectural focus. First, the organizational scope describes which part of the organization the EA function covers. For example, the EA function may cover the entire organization, it may focus on one business division, or it may even reside within one department. Second, there are four organizational levels the EA function may operate at, depending on its: (1) enterprise, (2) domain, (3) project, and (4) operational [24], [8]. Enterprise level is the highest organizational level, at which decision making aims at setting a long term strategic direction for the organization and achieving horizontal alignment between domains. At the domain level, decision making aims at setting domain specific objectives and optimizing the domain in order to achieve those objectives. A domain may be a specific business division, but may also be a generic functional domain that ranges over several divisions. For example, Customer Relationship Management (CRM) may be defined as a functional domain which provides a CRM generic service to several lines of business within one company. At the project level, decision making focuses on realization of the enterprise and domain level strategies by running change programs and projects that develop solutions. At the operational level, decision making aims at maximizing stability and continuity of the operational environment, such as systems, processes, and procedures. Third, the architectural focus of the EA function indicates which of the four aspect areas it covers: (1) Business structure and processes (B), (2) Information needs (I) of the business, (3) Information Systems (IS) delivering information services, and (4) Technical Infrastructure (TI) [25].

Based on its positioning and strategy, the *organizational structure* states whether the EA function is centralized, decentralized, or follows a federated model [23]. It gives an overview of the purpose, tasks and responsibilities of all functions, bodies and roles within the entire EA function.

The *operating model* describes the working of the EA function. In order to support EA decision making and ensure EA conformance, the EA delivery function may operate in three ways: (1) providing informal advice, (2) providing formal advice, or (3) acting as a formal gatekeeper. Providing *informal advice* is the most noncommittal approach in which architects provide advice and support, but have no means available to enforce EA products and stop projects that do not conform to those EA products.

When architects provide *formal advice*, management is obliged to examine their advice. Management is responsible for EA decision making, and must sign responsibility for potential consequences stated in the advice. If the EA delivery function is a *formal gatekeeper*, architects are mandated to take EA decisions and enforce EA products by stopping projects that deviate from EA products.

4.1.2 EA Governance

The EA *governance structure* describes how the EA functions, bodies and roles are integrated into the organization structure. For the governance bodies of the EA function it describes which functions and roles take part in those bodies. For example, the EA council at enterprise level typically has representatives of the various domains involved in EA decision making. There may also be an integrated project review committee comprised of architects and subject matter experts who validate project deliverables and operational changes. In line with the EA governance structure, the *responsibilities and authorizations* describe in detail the RACI-elements (Responsible, Accountable, Consulted, and Informed) of the EA functions, bodies, and roles.

The *governance processes* of the EA function describe how EA decision making is formalized and how EA products are approved and enforced. For example, it describes where in the project life cycle project deliverables must be validated on their EA conformance. It also describes how issues of non-conformance are handled through granting or rejecting escalations and waivers. The *rules and procedures* involved with EA decision making and EA conformance describe the forms, templates, guidelines, and criteria that apply regarding EA decision making and EA conformance (e.g., a waiver template or project validation criteria). When properly observed, procedures and rules enable transparent and consistent EA decision making and EA conformance, which is vital for the acceptance of the EA function's outcomes.

4.1.3 Collaboration and Communication

Relational integration is the voluntary and collaborative (informal) behavior of the various stakeholders of the EA function to clarify differences and solve problems in order to find integrative solutions [23]. Active stakeholder participation in the EA function is vital for its efficiency. Mechanisms to facilitate such relational integration are social networks (e.g., architecture community), and joint performance incentives.

Structural, transparent, and consistent *information exchange* between the various stakeholders of the EA function, with different functions and roles, is a critical success factor for the EA function. Information exchange involves both the communication of EA decision making and the reporting of the operational performance of the EA function – e.g., regarding the EA conformance of projects, or the functioning of the EA delivery function.

There should be an adequate *strategic dialogue* in order to properly facilitate EA decision making, such as identifying synergy opportunities, and resolving diverging perspectives and conflicts between various stakeholders within the EA function. A strategic dialogue involves exploring and debating ideas and issues outside formal EA decision making, incorporating various perspectives and views [23].

Shared learning among the stakeholders representing the various functions, roles, and bodies within the EA function allows a continuous learning and improvement cycle. Such a learning cycle is created by incorporating a feed-forward and feedback loop in the processes of the EA function. The feed-forward loop includes EA delivery providing pro-active support on applying EA products. The feedback loop enables stakeholders at project and operational level to share their practical experience with applying the EA products and suggesting improvements [8].

4.2   Part 2: The EA Delivery Function

In order to assess the EA delivery function's efficiency we created the model shown in Figure 2. From experience we learned that there is a gap between theory, which shows how well an idea or plan is described, and practice, which determines how well the idea is executed [16]. One of the reasons for this is the tendency of EA delivery functions to suffer from the ivory tower syndrome [9]. In our approach we assess the EA delivery function on both their levels of theory and practice.

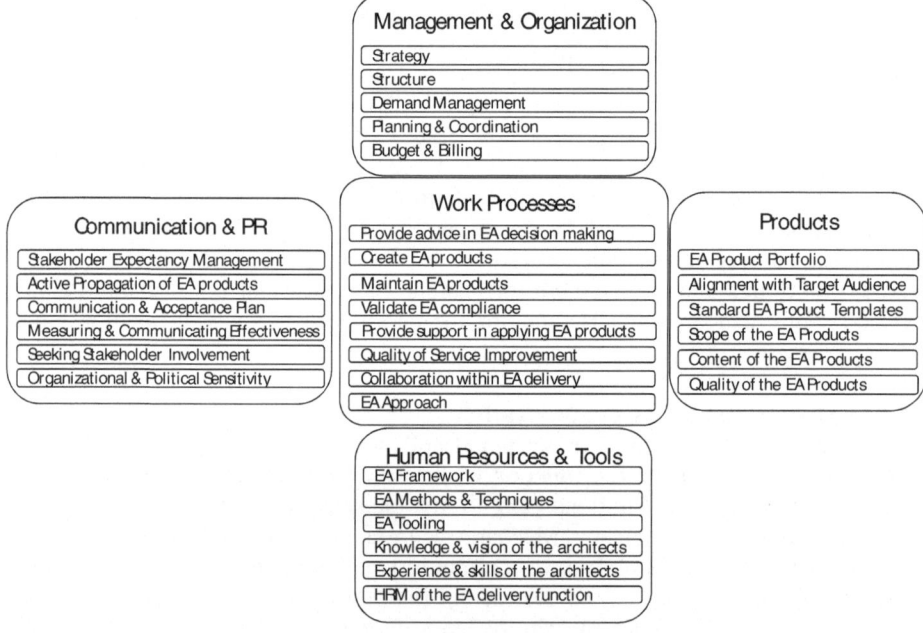

Fig. 2. EA delivery efficiency assessment topics

4.2.1   Management and Organization

The *strategy* of the EA delivery function describes its mission, strategic objectives, activities, and added value. It should be based on a stakeholder analysis, and must be aligned with the strategic goals of the organization and the entire EA function (see Section 4.1.1). The *structure* of the EA delivery function describes its internal structure – e.g., architect roles, task descriptions, architecture teams, etc. In order to ensure this structure is followed in practice, EA management must supervise whether the architects act according to the defined structure. *Demand management* requires the EA delivery function to know how much work of which type is requested by the stakeholders. For example, knowing the number of validations of project proposals for the coming month is essential, since the workload may fluctuate. Based on the demand, architecture management can perform *planning and coordination* in order to prioritize and divide the EA delivery activities over the architects. Finally, architecture management is responsible for the *budget and billing* of the EA delivery activities. Getting the required budget involves creating a business case that shows the benefit of having an EA function exceeds the costs.

4.2.2 Communication and Public Relations

The starting point for service delivery and communication of the EA delivery function should be the stakeholder expectations. *Stakeholder expectancy management* requires the EA delivery function to perform a stakeholder analysis in order to adapt its service provision to its stakeholder's expectations. Important to note is that these expectations should be in line with the goals as part of their formal role, and not with their personal and political goals. *Active propagation of EA products* is essential to keep the stakeholders up-to-date regarding the products and services of the EA delivery function. This activity should be part of an integral *communication and acceptance plan* which describes how the architects explain the purpose of EA, positively influence the attitude of the stakeholders towards the EA function, and get the stakeholders to comply with the EA products as second nature. *Measuring and communicating the effectiveness of EA* [26] is also part of communication & PR for the EA delivery function. Providing proof that the objectives with EA are being achieved should improve the acceptance and attitude of stakeholders towards EA. In addition to a communication plan, the EA delivery function should have a plan for *seeking stakeholder involvement*. This involves getting important change agents (e.g. senior management, highly respected subject matter experts, etc.) to positively influence others to accept and actively work together with the EA delivery function [27]. And finally, architects that are aware of and can cope with *organizational and political sensitivity* is essential in getting organizational acceptance and support for the EA delivery function.

4.2.3 Work Processes

As described in Section 2.2, the EA delivery function has 5 primary activities. First, *provide advice to support EA decision making* by the EA governance bodies. This involves determining what type of advice the EA decision makers require. How the decision makers would like that advice structured and presented to them so that they can comprehend the implications of their decision making. Second, *creation of EA products* involves gathering the essential EA requirements, devising solution alternatives, analyzing the consequences for each of these alternatives, guiding the decision makers in choosing the best alternative, and performing a scenario assessment of the chosen alternative to determine their behavior in case of a changing environment [28]. Third, *maintenance of EA products* includes processing changes to, or the retirement of existing EA products, clearly logging their status and version number throughout their life cycle. Fourth, *validating EA compliance* involves reviewing changes, implemented by projects and operational maintenance, on whether they conform to the EA products. Fifth, *provide support in applying EA products* entails pro-actively explaining their purpose, showing how to apply them, and providing feedback, hints and tips for applying them in practice.

All five activities described above should be incorporated into a standard *EA approach*. This approach should frequently be reviewed and improved, and should be integrated with other approaches within the organization – e.g., the project management approach. In order to allow *quality of service improvement*, the EA delivery function should perform efficiency assessments of the EA (delivery) function, combined with quality of service assessments – e.g., through a stakeholder satisfaction assessment. If the effectiveness or quality of service delivery is too low, the internal activities of the EA (delivery) function must be changed to improve the effectiveness or service quality level. Proper *collaboration within EA delivery* is essential for the architects to act as one team and communicate one message to the stakeholders. Therefore, the architects should have frequent meetings. Tools – e.g., an online architecture forum – or other collaboration instruments – e.g., an architecture community – also facilitate cooperation between architects (see Section 4.2.4).

4.2.4 Human Resources and Tools

The EA delivery function should have a standard *EA framework*, e.g., [25], [29] which provides the architects with a shared meta model to define the EA artifacts and their relations, and a shared terminology and common language (e.g., Zachman [29] or IAF [25]). EA methods and techniques provide the means to perform impact analyses, create models, and implement architectures. For example, TOGAF is a methodology which describes the process of creating an architecture [30]. The EA framework, as well as the *EA methods and techniques*, should be made and kept fit-for-purpose. Regarding tool support, the EA delivery function should use *EA tooling* for modeling and creating EA products, and performing impact analyses for decision making. A central EA knowledge base should be used to share EA products among architects to work with and reuse when possible. A publication tool should be used to publish EA products for stakeholders to read and use.

The human resources of the EA delivery function are its main asset. The *knowledge of the architects* should cover the specific aspect areas and organizational scope and levels the EA function focuses on (see Section 4.1.1). This knowledge is essential in order to create and maintain high quality EA products. Also, architects typically have a strategic advisory role. The *vision of the architects* on the business trends, technological innovations, and regulatory developments is essential in providing advice and creating EA products. Architects typically deal with conflicts between strategic decision making, institutionalized in EA products, and the realization of the strategy in programs, projects and operational environment. In order to be effective in their role as strategic advisor and safe guarder of EA conformance, the *experience and skills of the architects* is essential. Finally, to ensure the knowledge, vision and skills of the architects is sufficient, there should be proper *Human Resource Management (HRM) for the EA delivery function*. This includes competence management, based on profiles for the various roles within EA delivery function and an overall competence profile of entire EA delivery function. A professionalization program including recruitment policy, training program, coaching structure, and mechanisms for personal development plans should also be in place.

4.2.5 Products

The EA delivery function should have an *EA product portfolio* describing which types of EA products it delivers. Each product type in the portfolio should be aligned *with the target audience* (stakeholders of the EA function) based on their expectations concerning quality and content of the EA products, and their background knowledge. Therefore, there should be *standard EA product templates*, and the contents of the EA products should be predefined. Regarding The *scope of the EA products,* the EA delivery function should cover the aspect areas and organizational levels the EA function focuses on (see Section 4.1.1). Also, the *contents of the EA products* should include a description of the strategy, requirements, logical solution alternatives, physical solutions, transformation plan [25], and EA policies [2] for a specific domain or the entire organization, depending on the EA function's positioning.

To guarantee the *quality of the EA products*, the EA delivery function should have a quality mechanism in place. This quality mechanism involves describing the quality requirements of the EA products (e.g., Recognizable, Comprehensible, Relevant, Up-to-date, Consistent, Coherent, Accessible, Useful, Realistic, Pragmatic, Complete, Coherent), frequently performing EA product quality audits based on those requirements, and the improvement of the EA product portfolio and the products the EA delivery function produces based on the quality feedback.

5   Efficiency Assessment Approach

In order to conduct benchmark-quality EA function efficiency assessments, we use the Standard CMMI Appraisal Method for Process Improvement (SCAMPI) [31] tailored to the specific characteristics of NAOMI assessments. The SCAMPI approach consists of 3 phases: (1) plan and prepare the assessment, (2) conduct the assessment, and (3) report assessment results. Phase 1 consists of steps to create the assessment plan and prepare the assessment team. Phase 2 contains steps to conduct the assessment by preparing and performing evidence finding (through interviews and document study), validating these findings, and generating the assessment results. The last step in phase 2 includes creating a profile description of the EA function, an overview of the assessment findings, and an EA delivery efficiency profile using the NAOMI scoring questionnaires. The final phase consists of steps to present and archive the assessment results, including recommendations and an improvement plan.

Figure 3 provides a deliverable oriented overview of the NAOMI approach. An essential part of the assessment preparation is clearly defining the *assessment scope*

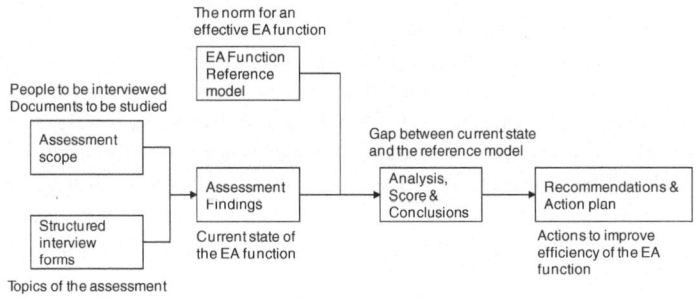

Fig. 3. Deliverable oriented overview of the NAOMI approach

based on the definition of the EA function (see Section 4.1.1). A clear assessment scope decreases the risk of losing focus, or running over time during the assessment. It also allows to clearly identify which stakeholders should be interviewed, and which documents should be studied in order to gather the essential *assessment findings*. A master interview form describes all topics of our EA function efficiency assessment model described in Sections 4.1 and 4.2. For each stakeholder the relevant interview topics described in the master interview form are copied into *structured interview forms* based on their roles and responsibilities regarding the EA function. The documents selected forehand, and identified during the interviews, provide objective data to cross check the subjective findings from the interviews.

During data gathering through interviews and document studying, it is key to periodically determine how many of all topics in the assessment model have been covered by the data gathered so far. Such an inventory allows you to better focus the coming interviews in order to ensure that all topics in the assessment model have been investigated. This is an essential condition to shift from data gathering to performing the analysis and determining the assessment score.

The assessment analysis involves describing the specific *current state of the EA function* being assessed, and the gap with the generic *EA function reference model* [8]. This gap *analysis* provides the key input for determining the efficiency score of the EA delivery function using the standard scoring questionnaire. For each topic in our EA delivery function efficiency assessment model (see Section 4.2), the scoring questionnaire contains several statements that describe the norm for each topic, based on our EA function reference model [8]. The assessors score the specific EA delivery function by indicating, on a 5 point Likert scale [32] (0=Strongly disagree to 5=Strongly agree), how well it achieves the norm. The scores of all assessors are combined into an efficiency profile consisting of a *score* for the five main topics in our assessment model (see Figure 4), and for each sub-topic as described in Section 4.2. The final step in an EA function efficiency assessment is to define *conclusions* and present *recommendations* and an *action plan* for improvement.

6 Case Study: Back-Office of a Large International Company

We conducted an assessment of the EA function within the back-office of a large international company, henceforth called company A. We conducted the assessment with a team of 6, consisting of one lead assessor, 4 assessors, and 1 scribe. We followed the approach we describe in Section 5.

The back-office is responsible for delivering operational IT services to and creating new IT solutions for the European front-office divisions of company A. At the time of the assessment, the back-office consists of 8 Lines of Business (LoB), four of which service the retail divisions and the other four the wholesale divisions of the front-office. The back-office of company A has one central technology division responsible for providing infrastructural services to the LoBs. The technology division is divided into a strategic and tactical Technology Office (TO), and an operational Technology Services (TS) department.

In order to assess the efficiency of the EA function within the back-office of company A, we conducted 49 interviews. 28 interviews were conducted with architects and architecture managers. The remaining 21 interviews were with stakeholders of the EA function (i.e., solution designers, project managers, program managers and directors, and various members of the TO management team). The interviews with the architects and architecture managers all lasted 1.5 hours. The interviews with the other members of the EA function lasted between 1 and 1.5 hours. We used fully structured interview forms for all 49 interviews to ensure all required topics were addressed. All interviews were performed by two people, one interviewer and one scribe. During the interview, the scribe took detailed minutes and the interviewer made personal notes. After the interview was finished, the scribe created a draft interview report based on the minutes. The interviewer checked this interview report with his/her personal notes and made required changed. Following, the all draft interview reports were sent to the interviewees for review. The interviewers finalized the interview reports based on the feedback of the interviewees.

In order to validate the findings from the interviews, the assessors analyzed 76 documents (e.g., EA products, EA governance and management documentation, and existing audit reports). The lead assessor facilitated several workshops with all assessors to consolidate all validated findings into one overall description of the entire EA function of company A's back-office. Based on this overall description, all assessors filled out the standard scoring questionnaire. These individual scores were consolidated into the final assessment scores.

6.1 EA Function Efficiency Assessment

6.1.1 Definition of the EA Function

The EA function we assessed is part of the TO department of the Technology division. The *organizational scope* of the EA function is the entire back-office organization of company A, and its *architectural scope* is limited to the Technical Infrastructure (TI) domain. The EA function does not cover the TS department and therefore only investigates the change activities of the projects run by the Technology division. The TI domain covers the entire scope of the back-office organization. Therefore, the EA function's *organization level* is at enterprise level. The TI domain is divided into 11 sub-domains, such as middleware, security, and networking.

Regarding the *strategy* of the EA function, its mission is to ensure the TI solutions implemented by TO are of high quality, and simplify the IT landscape. As part of the EA function, the EA delivery function aims to contribute to this mission by: (1) creating and maintaining TI policies and TI domain architectures, and (2) validating solutions whether they comply with them. The EA function is not involved in validating operational changes to the technical infrastructure implemented by the TS department on their conformance to the TI policies and architectures.

The *organizational structure* of the EA function consists of the *EA delivery* function with 11 infrastructure domain architects, 1 for each TI sub-domain, and one chief architect responsible for functionally managing the domain architects. Hierarchically, the domain architects are part of a resource pool of architects, managed by resource managers. The stakeholders (i.e., project managers, and infrastructure designers and engineers) involved with running projects make out the *EA conformance* function. There is no separate stakeholder group or body outside the EA delivery function formally responsible for *EA decision making*.

The *operating model* of the EA delivery function is to create TI policies and architectures, and performing solution validations without providing pro-active support to projects. The domain architects try to act as gatekeepers, stopping projects that do not conform to the TI policies and architectures. The chief architect makes the final decision whether a project is allowed to continue.

6.1.2 EA Governance
Regarding the *governance structure*, there are no clear definitions of *responsibilities and authorizations* for the functions, bodies and roles that are part of the EA function. For example, EA decision making about TI policies and architectures is unclear. There is no formal EA council to approve the EA products. Therefore, the status of many EA products is undefined. There is a TI policy committee that discusses new or changed TI policies. However, this committee has no representation of key stakeholders outside the EA delivery function, and is not mandated to formally approve or reject TI policies or other EA products.

The *governance processes* within the EA function lack clear *rules and procedures*. For example, most solution validations take place at a too late stage of the project life cycle. The first validation of a solution design typically takes place when the solution design has been created. In that stage, the project has already been given management approval to start up based on an initial business idea. Therefore, if the solution design of the project is not compliant with the TI policies and architectures, stopping the project is very hard. There is no formal escalation or waiver handling procedure to resolve conflicts or exceptions to TI policies. Also, when a validation outcome is overruled and the project is allowed to continue, this is not transparently communicated towards all stakeholders involved.

6.1.3 Collaboration and Communication
The EA function we assessed in organization A has practically no structural relational integration mechanisms to facilitate collaboration, communication, and shared learning between architects and the other roles involved in the EA function.

6.2 EA Delivery Function Efficiency Assessment

Based on the interviews with 28 members of the EA delivery function (architects and architecture managers) we created the efficiency profile shown in Figure 4.

Regarding *management and organization* of the EA delivery function, the gap between theory and practice is relatively large. There is a description of the strategy and structure of the EA delivery function, but the architects are not fully aware of them. There are means for operational management of the EA delivery function, such as work planning and coordination. There is no insight, however, in the demand for EA support from projects and programs. This makes it hard to plan the activities of the architects ahead. Management of the EA delivery function has many ideas and plans for improvement, but these are not being implemented yet.

When it comes to *communication and PR*, there is no integral communication plan. Most architects have reasonable understanding of the expectations of stakeholders regarding the service provision of the EA delivery function. However, most of the architects are not actively trying to involve stakeholders outside the EA delivery function to participate. The products the EA delivery function delivers are not actively communicated to the stakeholders. The products are statically published at various locations on the company intranet, making them hard to find for stakeholders.

The *working processes* of the EA delivery function may be characterized as bureaucratic and reactive. The architects provide little support in applying the TI policies in practice. Their way of working is highly individualized, and hardly formalized. There is little collaboration and communication between architects. This leads to conflicting EA products, advices, and project validation outcomes.

The EA delivery function also scores low on *human resources and tools*. Most architects are highly valued for their technical knowledge of the infrastructure domains they are responsible for. However, their soft skills need improvement. In practice there is no coaching or training structure to improve this. There is a standard EA framework available, but most architects ignore this framework and there are several other frameworks being used. The architects use a document sharing tool as a knowledge repository, but reuse of EA products or artifacts hardly takes place.

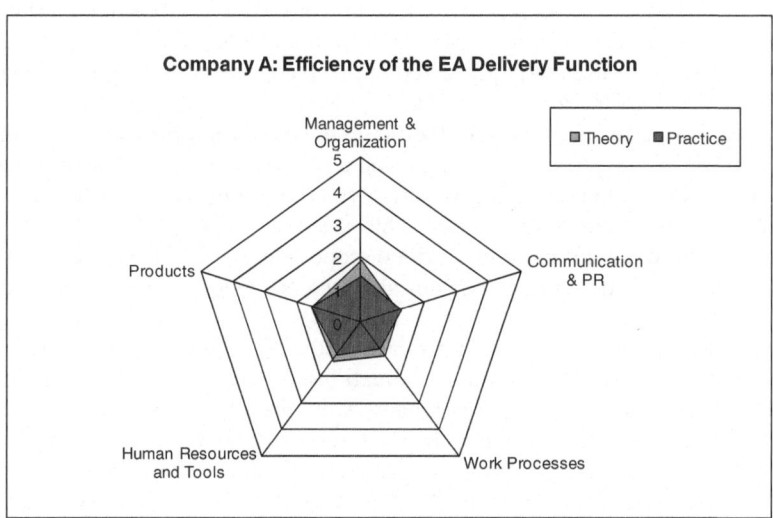

Fig. 4. Efficiency profile of the EA delivery function in Company A

Regarding the *products*, the total set of TI policies and architectures is ineffective. There are too many policies, and they are inconsistent. There is no standard EA product portfolio available, as well as no standard templates for EA products.

6.3 Conclusions and Recommendations

The main conclusion is that the EA function of organization A is hardly efficient. There are few elements to grasp on to in order to come with focused recommendations for improvement. We recommended organization A to implement an integral EA function, expanding the organizational scope to the entire back-office organization including the 8 LoBs. The architectural focus should also be expanded to include Business, Information, and Information Systems aspect areas. We recommended implementing a new EA function following a federated model, in which a central EA function creates, maintains, and enforces central EA products that set the boundaries for the 8 LoBs. Within these boundaries, the LoB specific EA functions may create, maintain, and enforce their own EA products. This provides the 8 specific EA functions with enough space to deal with LoB specific issues, but keeps the 8 LoBs aligned with the enterprise level strategy, architecture, and policies.

7 Discussion

7.1 NAOMI Model and Approach

The assessment model fits the situation of company A very well. We were able to come to a detailed and fully validated efficiency score on all topics in the assessment model. As a result of the assessment performed at company A, we were asked to help them implement a new EA function with a wider scope, covering the entire back-office of the large financial services company regarding its business, information, information systems, and technical infrastructure architectural areas.

A limitation of the case study presented in this paper is that its organizational scope is limited to the EA function responsible for the technical infrastructure of company A. However, we performed several other assessments with a previous version of our model (i.e. [16], [17]). These assessments had a wider organizational scope and architectural focus, and included the business, information, information systems, and technical infrastructure architectural areas. This previous version already seemed very applicable in these situations with a wider scope and focus. Therefore, we have no reason to believe this new version of our assessment model should not fit such situations as well.

Based on the findings from previous case studies (i.e. [16], [17]) we changed our model to be more flexible to fit various situations without having to be altered. As a result, the assessment performed at company A required no changes to the assessment model. Therefore, we have reason to believe our model is flexible enough to fit any situation at any organization. Also, we validated the external validity of the model, because all assessment topics seemed relevant, and we were able to determine an efficiency score for all topics in the model.

Company A did score quite low on some assessment topics because they were al-most non-existent – e.g., company A had almost no structural relational integration mechanisms to facilitate collaboration and coordination between architects and stake-holders. This does not mean that our assessment model was incorrect or over- complete. It means that company A had to improve these topics dramatically, because their inexistence led to low efficiency. Therefore, an important part of implementing the new EA function at company A involved setting up structural integration mechanisms to improve the collaboration and coordination between stakeholders and architects.

7.2 Lessons Learned from Practice

We discuss the lessons learned with applying NAOMI to assess the efficiency of the entire EA function (in Section 7.2.1) and the EA delivery function (in Section 7.2.2).

7.2.1 Efficiency of the Entire EA Function

We observed low EA effectiveness within many organizations, due to low efficiency of the EA function. We experienced that organizations typically put a lot of effort into increasing the efficiency of the EA delivery function, expecting this will result into higher EA effectiveness. However, this is often not the case. The EA delivery function is an important element of the EA function, but depends on the willingness of other functions (i.e., EA decision making and EA conformance), bodies (e.g., EA council) and roles (e.g., project managers and designers) to participate. We recommend that more effort is put into increasing the efficiency of the entire EA function. This is done by ensuring the three key ingredients for an efficient EA function – a clear EA function definition, efficient collaboration and communication between all stakeholders involved, and efficient EA governance (see Section 4.1) – are available.

From interviews with numerous EA function stakeholders (e.g., domain owners, project managers, designers) at various organizations, we learned they have high expectations regarding the efficiency and effectiveness of the EA function [17]. In practice, many organizations have not reached their full potential regarding efficiency and effectiveness of their EA function. The gap between the expectations and the degree in which these expectations are met typically results in low stakeholder satisfaction [18], [33]. Low satisfaction reduces the willingness of the stakeholders to actively participate and help increase EA function efficiency. We recommend frequently holding *EA stakeholder satisfaction surveys* to get insight into the stakeholder's expectations and their perception of how well these are met. Based on the results of these surveys, the points for improving EA function efficiency should be prioritized to increase EA stakeholder satisfaction and EA stakeholder involvement.

We experienced that assessing the efficiency of the EA function results in useful insights. Comparing the current state of an EA function to our EA function reference model allows identifying the strong points and points for improvement. However, we also found that we need to assess more than process quality and resource consumption in order to know whether improvements are really worth the investment. Since EA is a means to an end, organizations need to set the right strategic objectives for the EA function and *measure outcome effectiveness* in order to improve its efficiency. For example, one may have a highly efficient process to build life jackets made of concrete. Clearly, in this case process efficiency was not aligned to outcome effectiveness. Therefore, we recommend organizations to set clear goals with the EA function [34], define practical measurable effectiveness indicators, and improve EA function efficiency accordingly.

7.2.2 Efficiency of the EA Delivery Function

We observed that architects are typically highly knowledgeable, senior employees of organizations. They are valued for the (mostly technical) support they provide in decision making at various levels of the organization. However, their soft skills (communication and collaboration) and their attitude are often problematic.

We learned that architects, while validating projects and operational changes on their EA conformance, typically apply the governance rules quite strictly. When following the gate keeper model, the architects are supposed to apply them strictly. However, we found that architects often fail to provide proper feedback in case of a negative validation outcome. Not knowing why their project is stopped, or what to do in order to achieve EA conformance, project members get frustrated with EA. This eliminates their willingness to participate in the EA function. Also, we have observed that the attitude of EA delivery function members is typically reactive. This leads to frustration and misunderstanding with the other stakeholders of the EA function. For example, project managers want to know which rules and procedures they need to follow in order to be EA compliant. Architects typically fail to explain the EA governance rules and procedures before a project starts, as well as fall short in providing support on how to apply them during the course of a project. We recommend the EA delivery function to train their architects and change their way of working in order to provide *proactive support* to EA stakeholders.

Our observation is that architects seem to be more concerned with increasing the efficiency of EA as a means instead of the end. For example, architects at domain and project level are often little aware of the strategic objectives with EA or the effectiveness of the EA function. They are more interested in EA frameworks, tools, and technological or business innovations that help them create higher quality architectures. Also, we observed in practice that architects are typically quite concerned with delivering a high quality solution, and not so much with actually taking the responsibility for ensuring this solution is realized in practice. This partly explains their often internally oriented and reactive attitude towards EA stakeholders outside the EA delivery function. We recommend that architects become more *goal oriented*, by making them aware of the goals of the EA function, and making them responsible for achieving those goals.

7.3 Related Work

The Enterprise Architecture Management Maturity Framework (EAMMF) has been developed by the United States General Accounting Office [12]. The framework describes 31 core elements, which are descriptions of a practice or condition needed for an efficient EA delivery function. The framework associates each core element to one of five hierarchical maturity stages, and one of four types of management attributes, referred to as critical success attributes. Because it uses several maturity stages to assess a specific EA function, EAMMF assumes that typical patterns of core elements and management attributes apply to a specific maturity stage. To our experience, the development path of EA may differ for each organization [14]. Therefore, in specific situations these patterns may not apply. Also, linking specific elements and attributes to specific maturity levels makes it hard to perform an integral assessment on all relevant topics to identify improvement points.

Two models comparable to EAMMF are the Enterprise Architecture Maturity Model (EAMM) created by the National Association of State Chief Information Officers [11] and the Enterprise Architecture Capability Maturity Model (EACMM) created by the United States Department of Commerce [13]. Both models have a comparable background as the EAMMF. They include some elements of the entire EA function, but provide much less detail and do not refer to any guiding documents. Like EAMMF, these models also describe a hierarchy of 5 maturity stages.

Ross et al. describe a vast number of practices for building a mature EA function [4]. These management practices are divided into two categories: processes and roles. They do not provide a means to perform maturity assessments based on these management practices, plus these management practices are related to 4 maturity stages.

Van den Berg et al. provide a practical guide to building an EA capability, focusing mainly on the EA delivery function [6]. Besides many practical methods, techniques and tools, they provide a practical approach to assessing the maturity of the EA delivery function by applying a staged maturity model. Their approach does not provide a separate reference model of the EA function. Also, their approach is positioned as a self-assessment, which is likely to result in a biased outcome.

Bass et al. describe an approach for evaluating and improving the architecture competence of the EA delivery function [10]. The approach incorporates four models to assess: (1) the duties, skills and knowledge of architects, (2) the competences of the EA delivery function, (3) the cooperation between architecture teams, and (4) the learning cycle in the architecture design process. They provide a framework for building an assessment instrument, but do not provide one standard, ready-to-use assessment instrument as well as a separate EA function reference model.

8 Conclusions

In this article we present our assessment model to determine the efficiency of the EA function as part of our NAOMI approach. The model describes the topics on which the current state of an EA function is compared with our EA function reference model [8]. The assessment model also creates an efficiency profile of the EA delivery function using a standard scoring questionnaire. We used existing EA assessment models and practical experience gained while conducting assessments to create our model. In this article we use a case study to illustrate how our model works.

Our model provides insight in the current state for both the entire EA function as well as the EA delivery function. For the EA function, the model provides insight in the strategy, role and positioning of the EA decision making, EA delivery, and EA conformance functions. In addition, it aims at determining the formal governance structure and processes, as well as the informal communication and collaboration capabilities. In combination with our EA function reference model, this allows identifying points for improvement in order to build an integral and efficient EA function. The EA delivery function is scored on its management and organization, communication and PR, working processes, human resources and tools, and products in order to determine its efficiency. Having a separate efficiency profile for the EA delivery function enables the alignment of the EA delivery function to the requirements of the entire EA function. This is harder to accomplish with other existing EA assessment models because they provide one efficiency outcome for either only the EA delivery function, or the entire EA function. Existing models apply a staged maturity approach, describing a typical efficiency development path. This results in a less flexible approach which may not apply for some specific situations. The topics in our efficiency assessment model are not linked to maturity phases, which makes our model more flexible.

In practice, we found that few organizations have a truly efficient EA function. Many organizations focus on improving the efficiency of the EA delivery function, but we found that the scope of attention should be increased to the entire EA function. For example, more effort should be put into getting the EA stakeholders to actively participate in the EA function. We found that there is a large gap between the expectations of EA stakeholders regarding the efficiency and effectiveness of the EA function and the degree in which they are met. This is one of the reasons why the willingness of EA stakeholders to cooperate is low. We expect that a better fit between the EA function efficiency and the EA stakeholder expectations will improve their satisfaction and their willingness to participate. We are currently developing a standard EA stakeholder satisfaction survey, based on an exploratory study [17]. We also found that another essential input for efficiency improvement is the outcome effectiveness of the EA function. We are currently developing a measurement model to assess the effectiveness of the EA function [34].

References

1. Bharadwaj, A.S.: A resource-based perspective on information technology capability and firm performance: An empirical investigation. MIS Quarterly 24(1), 169–196 (2000)
2. Boh, W., Yellin, D.: Using Enterprise Architecture Standards in Managing Information Technology. Journal of Management Information Systems 23(3), 163–207 (2007)
3. Smolander, K., Päivärinta, T.: Practical Rationale for Describing Software Architecture: Beyond Programming-in-the-Large. In: Proceedings of 3rd Working IEEE/IFIP Conference on Software Architecture (WICSA3), pp. 113–126. Kluwer Academic Publishers, Dordrecht (2002)
4. Ross, J.W., Weill, P., Robertson, D.C.: Enterprise Architecture as Strategy – Creating a Foundation for Business Execution. Harvard Business School Press (2006)
5. Garlan, D.: Software architecture: a roadmap. In: Proceedings of the Conference on the Future of Software Engineering (ICSE 2000), pp. 91–101. ACM, New York (2000)
6. Van den Berg, M., Van Steenbergen, M.: Building an Enterprise Architecture Practice: Tools, Tips, Best Practices, Ready-to-use Insights. Springer, Heidelberg (2006)
7. Clerc, V., Lago, P., van Vliet, H.: The Architect's Mindset. In: Overhage, S., et al. (eds.) QoSA 2007. LNCS, vol. 4880, pp. 231–249. Springer, Heidelberg (2008)
8. Van der Raadt, B., Van Vliet, H.: Designing the EA Function. In: Becker, S., Plasil, F., Reussner, R. (eds.) QoSA 2008. LNCS, vol. 5281, pp. 103–118. Springer, Heidelberg (2008)
9. Kruchten, P.: The Software Architect. In: Donohoe, P. (ed.) Software Architecture (WICSA1), pp. 565–583. Kluwer Academic Publishers, Dordrecht (1999)
10. Bass, L., Clements, P., Kazman, R., Klein, M.: Models for Evaluating and Improving Architecture Competence. Technical Report, CMU/SEI-2008-TR-006 (2008)
11. NASCIO: NASCIO Enterprise Architecture Maturity Model, Version 1.3 (December 2003),
    http://www.nascio.org/publications/documents/nascio-eamm.pdf
12. Hite, R.: Information technology: A framework for assessing and improving enterprise architecture management (version 1.1). White Paper. United States General Accounting Office, Washington, DC (2003), http://www.gao.gov/new.items/d03584g.pdf
13. US DoC: Enterprise Architecture Capability Maturity Model, Enterprise Architecture Program Support. United States Department of Commerce., Version: 1.2 (December 2007)
14. Van der Raadt, B., Soetendal, J., Perdeck, M., Van Vliet, H.: Polyphony in Architecture. In: Proceedings 26th International Conference on Software Engineering (ICSE 2004), pp. 533–542. IEEE Computer Society Press, Los Alamitos (2004)
15. Van der Raadt, B., Hoorn, J.F., Van Vliet, H.: Alignment and Maturity Are Siblings in Architecture Assessment. In: Pastor, ., Falc o e Cunha, J. (eds.) CAiSE 2005. LNCS, vol. 3520, pp. 357–371. Springer, Heidelberg (2005)

16. Van der Raadt, B., Slot, R., Van Vliet, H.: Experience Report: Assessing a Global Finan-
    cial Services Company on its Enterprise Architecture Effectiveness Using NAOMI. In:
    Proceedings of the 40th Annual Hawaii international Conference on System Sciences
    (HICSS 2007), p. 218. IEEE Computer Society, Washington (2007)
17. Van der Raadt, B., Schouten, S., Van Vlict, H.: Stakeholder Perception of EA Function
    Performance. In: Morrison, R., Balasubramaniam, D., Falkner, K. (eds.) ECSA 2008.
    LNCS, vol. 5292, pp. 19–34. Springer, Heidelberg (2008)
18. Hoorn, J.F.: Software Requirements: Update, Upgrade, Redesign. Towards a Theory of
    Requirements Change. Ordina, Nieuwegein, NL (2006)
19. Sherehiy, B., Karwowski, W., Layer, J.K.: A review of enterprise agility: Concepts,
    frameworks and attributes. International Journal of Industrial Ergonomics, 445–460 (2007)
20. Luftman, J.: Assessing IT/Business Alignment. Information Systems Management 20, 9–
    15 (2003)
21. Cameron, K.S., Whetten, D.A.: Organizational Effectiveness and Quality: The Second
    Generation. In: Higher education: Handbook of Theory and Research, vol. XI. Agathon
    Press, New York (1996)
22. Ross, J.W., Beath, C., Goodhue, D.L.: Develop long-term competitiveness Through IT as-
    sets. Sloan Management Review 38(1), 31–45 (Fall 1996)
23. Peterson, R.: Crafting Information Technology Governance. Information Systems Man-
    agement 21(4), 7–22 (2004)
24. Pulkkinen, M.: Systemic Management of Architectural Decisions in Enterprise Architec-
    ture Planning. Four Dimensions and Three Abstraction Levels. In: Proceedings of the 39th
    Annual Hawaii International Conference on System Sciences (HICSS 2006), p. 179a.
    IEEE Computer Society, Washington (2006)
25. Mulholland, A., Macaulay, A.L.: Architecture and the Integrated Architecture Framework.
    Capgemini (2006),
    http://www.capgemini.com/services/soa/ent_architecture/iaf/
26. Cane, S., McCarthy, R.: Measuring the Impact of Enterprise Architecture. Issues in Infor-
    mation Systems VIII(2) (2007)
27. Kotter, J.P.: Leading Change: An Action Plan from the World's Foremost Expert on Busi-
    ness Leadership. Harvard Business School Press, Boston (1996)
28. Simonsson, M., Lindström, Å., Johnson, P., Nordström, L., Grundbäck, J., Wijnbladh, O.:
    Scenario-Based Evaluation of Enterprise Architecture - A Top-Down Approach for CIO
    Decision-Making. In: Proceedings of the International Conference on Enterprise Informa-
    tion Systems (ICEIS 2005), pp. 130–137 (May 2005)
29. Zachman, J.A.: A Framework for Information Systems Architecture. IBM Systems Jour-
    nal 26(3). IBM Publication G321-5298 (1987)
30. Open Group: TOGAF (The Open Group Architecture Framework) version 8.1.1 ('The
    Book') (2007),
    http://www.opengroup.org/bookstore/catalog/g063v.htm
31. SEI: SCAMPI A, Version 1.2: Method Definition Document. CMU/SEI-2006-HB-002
    (August 2006), http://www.sei.cmu.edu/publications/documents/
    06.reports/06hb002.html
32. Likert, R.: A Technique for the Measurement of Attitudes. Archives of Psychology 140,
    1–55 (1932)
33. Boster, M., Liu, S., Thomas, R.: Getting the Most from Your Enterprise Architecture. IT
    Professional 2(4), 43–50 (2000)
34. Van der Raadt, B., Bonnet, M., De Bruijne, M., Van den Berg, J., Van Vliet, H.: Effec-
    tiveness of Enterprise Architecture. In: Fifth International Conference on the Quality of
    Software Architectures (submitted, 2009)

Business Value of Solution Architecture

Raymond Slot[1], Guido Dedene[2], and Rik Maes[3]

[1] Capgemini, Utrecht, Netherlands
raymond.slot@capgemini.com
[2] University of Amsterdam, Netherlands
guido.dedene@econ.kuleuven.be
[3] University of Amsterdam, Netherlands
maestro@uva.nl

Abstract. The theory and especially the practice of IT architecture have been developed quite vigorously the last years. However, hardly any quantitative data about the value of IT architecture is available. This paper presents the results of a study, which measures the value of IT solution architecture for software development projects. The study identifies ten architecture-related project-variables and correlates these with eight project success variables. Statistical analysis of 49 IT projects shows that the use of solution architecture is correlated with decreased budget and time overrun, increased reliability of project planning and increased customer satisfaction. The results of the study indicate that IT usage of solution architecture for custom software development projects leads to better project results. Also the limitations of the study are discussed.

Keywords: Business Value, Solution Architecture, IT project success.

1 Solution Architecture

1.1 Introduction

The theory and especially the practice of IT architecture have been developed quite vigorously the last years. International and national standardization organizations, such as *The Open Group* [1] and in the Netherlands the *Telematica Instituut* [2] are working on standardization of business and IT architecture and the effects of these efforts are reaching the end users. Various IT organizations, such as Capgemini, have developed their own architecture framework and are using it in the market [3].

Considering the activities that take place in the business and IT architecture world, it is surprising that the business case for these activities is for a large part nonexistent. There is little research done to quantify, in financial terms, the value of architecture.

The main subject of this article is to quantify, in financial terms, the value of solution architecture for organizations. Organizations invest in solution architecture. These investments include training of architects, development of architectures and implementation of architecture processes. Is the spending of this money justified? Approaches to information economics [4] and [5]do not include the effects of investing in business and IT architecture. The key-question this article addresses is "How to quantify the value of IT solution architecture for an organization?"

E. Proper, F. Harmsen, and J.L.G. Dietz (Eds.): PRET 2009, LNBIP 28, pp. 84–108, 2009.

1.2 Required Disciplines for IT Projects

In the literature, project management, analysis & design and software development & testing, attract a lot of attention and many methods and approaches have been devised for these activities. For instance, for a discussion on project management see Kerzner [6] and PRINCE 2 [7]. Analysis & Design and Software Development & Testing are described in various development methods, among which RUP [8] and DSDM [9] are well known. In addition, CMMI [10] defines stages and maturity levels for (software) development processes.

1.3 Role of Solutions Architecture

None of these approaches recognizes explicitly the role of solution architecture, although the DSDM Consortium has published a paper on the relationship between TOGAF [1] and DSDM [11]. See for another initiative in this direction the Enterprise Unified Process [12].

According to Piselo [13], about one third of custom software development projects fail, about half of the projects is late, over budget, or has reduced functionality and only one sixth of the projects is delivered on time, within budget and according to specification. Considering the lack of attention of the major software development methodologies for architecture, one could assert that this is one of the reasons for this poor performance. In this paper, we will study the effects of solution architecture on projects. Concretely, we will test the hypothesis whether the success of software development projects is correlated with the usage of solution architecture.

1.4 Development under Architecture

Enterprise architecture sets standards and guidelines, based on strategy, for the structuring of the organization. The enterprise architecture is implemented by many projects, each implementing its own part of the total design. The approach where project objectives are also determined by enterprise architecture objectives is called development under architecture. Wagter [14] formulates this as follows: "Development under architecture realizes concrete business goals within the desired time frame, at the desired quality levels and at acceptable costs. [..] When a project is developed under architecture, the project starts with a so-called Project Start Architecture [(PSA)]. A [PSA] is a translation of the overarching [enterprise] architecture principles and models to rules and guidelines tailored to the project. This provides the practical rules, standards and guidelines used by the project. Also, project design choices that influence other projects are described in the [PSA]." Based on this, we define "development under architecture" as follows:

A project is developed under architecture if standards, rules and guidelines of the enterprise architecture are incorporated in the scope of the project, and these are tailored towards the objectives of the project, as described in a solution architecture document. Furthermore, the solution architecture describes how the software built by the project should interact with its environment.

1.5 Project Success

An IT project can be considered as a process with a number of inputs (called project variables) and outputs (success variables). The approach to measure project and success variables we have chosen is derived from Wohlin and Andrews. [15]. They state: "If ▨▨, ▨▨▨, ▨ ▨▨, ▨ ▨▨, are the project variables and ▨▨, ▨▨▨, ▨ ▨▨, ▨ ▨▨, are the success variables, then the objective is to identify which project variables are good estimators for which success variables. Project variables describe key drivers and characteristics of the software project and can be measured (or estimated) before the project starts [or during project execution]. Success variables are measured when the project is completed." Figure 1 illustrates this.

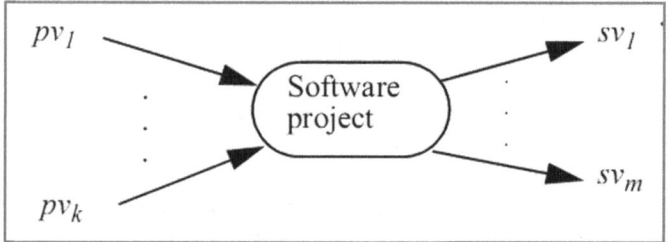

Fig. 1. Project and success variables (from Wohlin et al.)

Examples of project variables are: quality of the requirements, experience of the project manager, quality of the project architecture, etc. An example of a success variable is budget overrun.

Based on this model, we can define the effects of solution architecture on project success as follows:

Software development projects are more successful if we can identify solution-architecture related project variables, which are positively correlated with one or more success variables.

We selected ten architectural project variables and six success variables and we correlated each of the project variables with every success variable. For every significant correlation, the size of the effect will be calculated.

1.6 Project Variables

Jurgens [16] identified about 80 project variables or input variables to a project. He identified several categories of project variables: solution architecture-related variables, process-related variables, functional-requirement related variables, capability-related variables and IT-related variables. We selected the following architecture-related project variables:

Table 1. Overview of architecture-related project variables

Nr	Architecture-related Project Variable	Rationale for inclusion
1	Involvement of an architect in the technical budget calculation for the project	The technical budget is the budget for which the project should be built. Setting a technical prize includes considering various (technical) factors and price drivers, which is the expertise of the architect.
2	The general experience of the architect, who creates the project architecture, as reflected in the certification level of the architect.	Experienced architects have more insight oversight then less experienced architects.
3	The certification level of the architect, should match the complexity of the project.	Projects come in different sizes therefore and projects that are more complex should be linked to the more experienced architects.
4	The specific experience of the architect on the subject of the project.	It is probably advantageous for an architect to have experience with the specific topics of the project.
5	Quality of the solution architecture	The solution architecture is the guideline for the project.
6	Quality customer's domain architecture	The customer's domain architecture provides guidelines for the solution architecture.
7	Quality of the customers enterprise architecture	The customer's enterprise architecture is the guideline for the underlying domain and solution architecture.
8	Quality of the customers architecture governance process	A high-quality architecture governance process helps the project made the right decisions.
9	Presence of a controlling architect during the execution of the project	A controlling architect checks whether the project keeps itself to the solution architecture.
10	Compliancy testing between architecture and project during execution	If the compliancy is checked of project deliverables and the project as picture, then, any discrepancies between the two are known

1.7 Success Variables

For a selection of the success variables, we refer to Wideman [17]. He describes four dimensions of project success.

1. Internal Project Objectives (efficiency during the project)
2. Benefit to Customer (effectiveness in the short term)
3. Direct Contribution (in the medium term)
4. Future Opportunity (in the long term)

He gives the following key-questions and success variables for each of these dimensions.

Table 2. Dimensions, key questions and factors for project success according to Wideman

Dimension	Key-questions	Success variable
1. Internal	• How successful was the project team in meeting its schedule objectives? • How successful was the project team in meeting its budget objectives? • How successful was the project team in managing any other resource constraints?	• Meeting schedule • Within budget • Other resource constraints met

Table 2. (*Continued*)

Dimension	Key-questions	Success variable
2. Benefit to Customer	• Did the product meet its specified requirements of functional performance and technical standards? • What was the project's impact on the customer, and what did the customer gain? • Does the customer actually use the product, and are they satisfied with it? • Does the project's product fulfill the customer's needs, and/or solve the problem?	• Meeting functional performance • Meeting technical specifications & standards • Favorable impact on customer, customer's gain • Fulfilling customer's needs • Solving a customer's problem • Customer is using product • Customer expresses satisfaction
3. Direct Contribution	• Has the new or modified product become an immediate business and/or commercial success, has it enhanced immediate revenue and profits? • Has it created a larger market share?	• Immediate business and/or commercial success • Immediate revenue and profits enhanced • Larger market share generated
4. Future Opportunity	• Has the project created new opportunities for the future, has it contributed to positioning the organization consistent with its vision, goals? • Has it created a new market or new product potential, or assisted in developing a new technology? • Has it contributed additional capabilities or competencies to the organization?	• Will create new opportunities for future • Will position customer competitively • Will create new market • Will assist in developing new technology • Has, or will, add capabilities and competencies

Not all of the information for the success variables mentioned by Wideman where available for our survey. We were able to collect information about the following six success variables:

Table 3. Overview of success variables

No	Variable	Definition
A	Budget	Percentage under run or overrun for the project. We compare the actual project cost to the original project planning.
B	Time	Percentage under run or overrun for the project time. We compare the actual timeframe with the original, planned timeframe.
C	Customer Satisfaction	Customer's satisfaction assessment of project execution and result.
D	Percentage Delivered	The percentage of the intended results that are actually delivered by the project.
E	Functional Fit	The match between the required and delivered functionality; is the functionality delivered by the project in accordance with the planned functionality?
F	Technical Fit	The match between the required and delivered non-functional characteristics; is security, availability, performance, etc. of the delivered result is accordance with the planned characteristics?

2   Case Study Description

2.1   Objective and Approach

The objective of the case study is to test the hypothesis that software development projects are more successful when developed under architecture. This hypothesis is tested by correlating architecture-related project variables with project success variables. The hypothesis can be confirmed if we find significant correlations between architecture-related project variables and success variables.

2.2 Description of the Projects

Forty-nine projects were included in the study. These were all IT projects where software was developed, based on tailor-made specifications of customers. About half of the projects developed software for companies in the financial sector; the others are from industrial and governmental organizations. The types of projects are: transformation projects[1], merger and acquisition projects[2], single function integration projects[3] and lifetime extension projects[4].

2.2.1 Project Size
The project size was between € 50K and € 2,5M. The average project size is about € 700K, while the median size is about € 350K. See Figure 2 and Table 4 for an overview of the key figures of the projects within the survey scope.

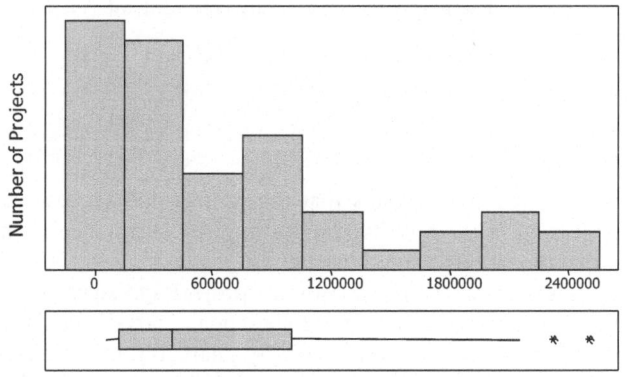

Project budget in Euros

Fig. 2. Histogram and box plot of project size

Table 4. Key figures of the size of the surveyed projects

Characteristic	Value
Number of projects	49
Average project size	€ 695.000
Minimum project size	€ 50.800
First quartile	€ 125.000
Median	€ 349.000
Third quartile	€ 939.000
Maximum project size	€ 2.500.000

---

[1] Cross functional projects, such as CRM implementations.
[2] Rationalization or consolidation projects, shared service centre.
[3] New general ledger or HRM system.
[4] Web enabling of legacy, application integration.

2.3 The Null-Hypothesis

Based upon the definition of the success variables (see Table 3), the following null-hypothesis statements are formulated.

Table 5. Null-hypothesis statements

No	H_0 statement
I	Usage of solution architecture is not significantly correlated with the expected value of the actual project budget (as percentage of the planned budget).
II	Usage of solution architecture is not significantly correlated with the variance of the actual project budget (as percentage of the planned budget).
III	Usage of solution architecture is not significantly correlated with the expected value of the actual project timeframe (as percentage of the original timeframe).
IV	Usage of solution architecture is not significantly correlated with the variance of the actual project timeframe (as percentage of the original timeframe).
V	Usage of solution architecture is not significantly correlated with the expected value of customer satisfaction.
VI	Usage of solution architecture is not significantly correlated with the expected value of the percentage delivered.
VII	Usage of solution architecture is not significantly correlated with the expected value of the functional fit.
VIII	Usage of solution architecture is not significantly correlated with the expected value of the technical fit.

Remarks

1. Note that the null-hypothesis states that the project and success variables are *not* correlated. Rejection of the null hypothesis implies acceptance of the alternate hypothesis, i.e. that the variables are correlated.
2. The analysis identifies *correlations* between project and success variables and not *causal relationships*, because the type of statistical analysis that is used, is not able to identify causal relationships. However, by analyzing these correlations, we are often able to give meaning to the correlation and describe a causal mechanism that may be underlying the correlation.
3. For both the budget and time success variables, we have defined two H_0-hypothesis statements. The budget and time success variables are tested both with regard to the expected value (statements I and III) and the standard deviation (statements II and IV). The other success variables are tested only for expected value. The reason to test budget and time for standard deviation is that architecture may be correlated with an increase of reliability of the planned budget and time. A lower value of the standard deviation indicates less variance in the outcomes of the projects and thus a higher reliability of the planned figures.

2.4 Measurement Setup

The approach we have chosen to collect the required information is by means of interviews with the project manager. One of our main points of attention was to ensure optimal reliability of the basic information, which was collected using the interviews. The following procedure was observed, to minimize ambiguities and errors in the answering of the questions:

1. Carefully formulate the questions and answers, to make them as unambiguous as possible. We first set-up a test questionnaire, used this questionnaire several times and then define the final questionnaire based on the experiences of the test interviews.
2. Setting up guidelines how to interpret the questions, especially in situations where the answers were not clear-cut. The interviewers used experiences from earlier interviews in later interviews.
3. Analyze the answers for specific patterns or outliers, which could indicate for misinterpretations or ambiguities questions. Use these experiences to revise the instructions for the interviewers or/and to adapt the formulation of the question and answers.
4. The interviewers were trained in interpreting the answers to questions as univocally as possible, by organizing discussions between them about the interpretation of questions.
5. Check the answers of the interviewees where possible by independent means. Some answers could be checked by information from financial systems, others by crosschecking it with other people who work on the same project.

This procedure delivered 49 filled-in questionnaires.

3   Case Study Results

3.1   Summary of Results

The table shows the correlations for project and success variables that test significantly – the probability value (p-value) of the test is equal of smaller than the significance level, for a chosen significance level of 5%. P-values that are not significant are replaced by a dash.

From the table can be concluded that all H_0 statements are rejected, except for H_0-statement VII. Project variables 2 and 9 are not significantly correlated to a success variable; the other project variables are all significantly correlated with at least one of the success variables.

Table 6. Overview results null-hypothesis testing

			H_0 Statement							
			I	II	III	IV	V	VI	VII	VIII
Project Variables	1	Technical Calculation	-	0,2%	-	-	-	-	-	-
	2	Certification Architect	-	-	-	-	-	-	-	-
	3	Certification w.r.t. Project	-	-	-	-	0,0%	-	-	-
	4	Specific Experience Architect	-	-	-	-	5,0%	1,5%	-	-
	5	Project Architecture	-	2,4%	0,2%	-	0,8%	0,2%	-	0,3%
	6	Domain Architecture	-	-	3,6%	-	1,9%	0,6%	-	-
	7	Enterprise Architecture	-	-	1,8%	3,5%	0,1%	2,6%	-	-
	8	Architecture Governance	0,3%	-	-	-	-	1,8%	-	-
	9	Controlling Architect	-	-	-	-	-	-	-	-
	10	Architecture Compliancy	-	-	1,0%	-	-	-	-	-

3.2 Interpretation of the Findings

In the following paragraphs, the findings are interpreted and explained. The interpretation follows a standard structure:

1. *Statement* – The formulation of the H_0 statement.
2. *Finding* – The actual findings from the analysis.
3. *Conclusion* – Conclusions that can be drawn from the findings.
4. *Significance* – Significance level. This level is equal or below the significant threshold of 5%.
5. *Interpretation* – Interpretation of the findings and the conclusion, which may provide additional reasoning or foundations for the conclusion.
6. *Consequences* – The size of the effect is explained, in terms of the effect on the success variable.

3.3 H_0 Statement I – Expected Value of Budget Overrun

3.3.1 H_0 Statement
Usage of solution architecture is not significantly correlated with the expected value of the actual project budget (as percentage of the planned budget).

Finding
Project variable 8 (*Quality of the customer's architecture governance process*) tests significantly. The other variables are non-significant. H_0 statement I is rejected.

Conclusion
The presence of an architecture governance process (either fully functional or limited in scope and responsibilities) is significantly correlated with a lower expected value of budget overrun, compared to a situation where there is no architecture governance process in the customer's organization present. The difference in expected value is 19%.

Significance
P = 0,3%

Interpretation
The presence of an architecture governance process implies that the organization is working with architecture and, therefore, is using solution architecture and higher-level architectures. The reverse situation is not necessarily the case; an organization may be defining solution architectures without having an architecture governance process. This finding shows that the presence of an architectural governance process has its own additional value.

Consequences
The average project size is € 700.000. A decrease the overrun with 19% will save on average € 130K per project, or about € 6M for the 49 projects that we have examined.

3.4 H₀ Statement II – Variance of Budget Overrun

H₀ Statement
Usage of solution architecture is not significantly correlated with the variance of the actual project budget (as percentage of the planned budget).

Finding
Project variable 1 (*Presence of architect during calculation of the technical price*) and project variable 5 (*Quality of the solution architecture*) tests significantly. The other variables are non-significant. H₀ statement II is rejected.

Conclusion
The presence of an architect during the calculation of the technical price is significantly correlated with a lower variance of the actual project budget, compared to a situation when there is no architect present during technical price calculation. The difference in the standard deviation of the project budget overrun percentage is 21 (13 versus 34).

The presence of a high-quality project architecture is significantly correlated with a lower variance of the actual project budget, compared to a situation when there is only a medium or poor quality or no project architecture present. The difference in the standard deviation is 18 (13 versus 31).

Significance
$P = 0,8\%$ (variable 1)
$P = 2,4\%$ (variable 5)

Interpretation
Presence of an architect during the calculation of the planned cost and the quality of the project architecture is correlated with an increase the reliability of the cost planning significantly. Reduction of variance is a major goal of the Six Sigma methodology [18]. When process variance is reduced, then the process becomes more predictable and overrun decreases. A major problem with custom software development is the lack of predictability of the actual cost. Both project variables *Presence of an architect during the calculation of the technical price* and *High-quality solution architecture* are correlated with a significantly improved reliability of the project budget planning.

Consequences
Piselo [13] states that only 16% of custom software development projects deliver their results according to plan. If the process variance is reduced, then this improves the process quality. For instance, we can calculate that only 13% of the projects with a high-quality solution architecture have more than 20% overrun, versus 38% of the projects with the medium or low quality solution architecture.

3.5 H₀ Statement III – Expected Value of Project Timeframe

H₀ Statement
Usage of solution architecture is not significantly correlated with the expected value of the actual project timeframe (as percentage of the original timeframe).

Finding
Project variables 5, 6, 7 and 10 (*Quality of the project architecture, Quality of the domain architecture, Quality of the enterprise architecture* and *Architecture compliancy testing*) test significantly. The other variables are non-significant. H_0 statement II is rejected.

Conclusion
Usage of solution architecture is correlated with a significant decrease in time overrun for projects. Four of the 10 project variables test significantly, which makes the project timeframe one of the success variables that correlate with multiple aspects of the use of architecture.

The presence of a high-quality project architecture correlates with a decrease in time overrun of the project, compared to a situation where there is a medium or poor quality project architecture present. The difference in overrun is 55% (71% overrun versus 16% overrun).

The presence of a high-quality domain architecture correlates with a decrease in time overrun of the project, compared with situation where there is medium or poor quality domain architecture present. The difference in overrun is 44% (49% versus 5% overrun).

The presence of a high-quality enterprise architecture correlates with a decrease in time overrun of the project, compared with situation where there is medium or poor quality enterprise architecture present. The difference in overrun is 46% (51% versus 5%).

The presence of an informal architecture compliance testing procedure correlates with a decrease in time overrun of the project, compared to the situation where there was no compliancy testing between architecture design and implementation. The difference in overrun is 56% (66% versus 10%).

Significance
P = 1,9% (variable 5)
P = 3,6% (variable 6)
P = 1,8% (variable 7)
P = 1,0% (variable 10)

Interpretation
It is interesting to note that four of the ten project variables correlate with the success variable. Probably, the same effect is measured multiple times, but from different angles. For instance, presence of enterprise architecture and the presence of the domain architecture denote probably the same type of architectural maturity of the customer's organization and both project variables may be an indication for a common underlying cause. Further indication of this is that variable 6 and variable 7 have almost the same expected values for time overrun. To understand this result more fully, it is necessary to analyze the interaction between project variables. However, the survey size is too limited to perform this type of analysis (see § *4.1*). Consequently, we have to limit ourselves to the supposition that interaction between project variables plays a major role in this result, without being able to quantify this interaction.

Overall, we can conclude that application of solution architecture is correlated with a substantial decrease in project time overrun.

Consequences
The average actual project timeframe for the projects that we have examined is one year – which includes on average 40% overrun. Consequently, application of architecture is correlated with a decrease of average project time of about four months.

3.6 H_0 Statement IV – Variance of Project Timeframe

H_0 Statement
Usage of solution architecture is not significantly correlated with the variance of the actual project timeframe (as percentage of the original timeframe).

Finding
Project variable 8 (*Quality of the enterprise architecture*) tests significantly. The other variables are non-significant. H_0 statement IV is rejected.

Conclusion
The presence of a high-quality enterprise architecture correlates significantly with a decrease of variance in the actual project timeframe, compared to a situation where there is medium or low quality enterprise architecture or no EA. The difference in the standard deviation of the percentage of the project timeframe overrun is 108 (115 versus 7).

Significance
P = 3,5%

Interpretation
The interpretation of this result is not very clear, because the difference in the standard deviation is quite large and the question is why we do not measure a correlation for domain and project architecture. Also, the sample size for one of the answers (answer 1 – *high-quality enterprise architecture*) is rather small – only 8 projects. The p-value for domain architecture is 11%, which could indicate a trend. However, the p-value for project architecture is 74%, which is nowhere significant. A possible explanation could be provided by the overall process maturity level of the organization. Higher process maturity may reflect itself in a high-quality enterprise architecture and this may influence the variance of project timeframe. We do not have information on process maturity of the organizations that are involved in the survey and, therefore, there is no way we can verify this theory. We suspect that this result could be spurious. Subsequent research may clarify this finding.

3.7 H_0 Statement V – Customer Satisfaction

H_0 Statement
Usage of solution architecture is not significantly correlated with the expected value of customer satisfaction.

Finding
Project variables 3, 4, 5, 6 and 7 (*Match of certification level of the architect to the level of the project, Specific experience of the architect, Quality of the project architecture, Quality of the domain architecture* and *Quality of the enterprise architecture*) test significantly. Project variables 2 and 8 (*Certification level of the architect* and *Quality of the customer's architecture governance process*) are close. H_0 statement V is rejected.

Conclusion
Usage of solution architecture is correlated with a significant increase in customer satisfaction. Five of the ten project variables test significantly, which makes customer satisfaction one of the success variables that correlates with multiple aspects of the use of architecture.

Matching the level of the architect with the level of the requirement correlates significantly with an increase of customer satisfaction, compared to a situation where the certification level of the architect was under project level. The difference is a customer satisfaction score of 4,1 versus 2,8 (on a scale of 1 to 5). This one project variable explains 51% of the total variance in the score.

Broad experience of the architect with the type of engagement correlates significantly with an increase of customer satisfaction, compared to a situation where the architect has only some experience with the type of engagement. The difference is a customer satisfaction score of 4,0 versus 3,6. This project variable explains 8,5% of the variance in the customer satisfaction score.

The presence of a high-quality project architecture correlates significantly with an increase of customer satisfaction, compared to a medium or low quality or no project architecture. The difference is a customer satisfaction score of 4,1 versus 3,5. This project variable explains 16,8% of the total variance of the customer satisfaction score.

The quality of the domain architecture correlates significantly with an increase of customer satisfaction. The customer satisfaction score is 4,2, 3,8 and 3,4 for respectively a high-quality, medium quality or low quality domain architecture. This project variable explains 12,5% of the total variance of the customer satisfaction score.

The quality of the enterprise architecture correlates significantly with an increase of customer satisfaction. The customer satisfaction score is 4,4, 3,9 and 3,4 for respectively a high-quality, medium quality or low quality enterprise architecture. This project variable explains 24,3% of the total variance of the customer satisfaction score.

Significance
P = 0,0% (variable 3)
P = 5,0% (variable 4)
P = 0,8% (variable 5)
P = 1,9% (variable 6)
P = 0,1% (variable 7)

Interpretation
Our supposition is that Customer Satisfaction score is the outcome of the comparison between the expectation of the customer and the actual outcomes of the project. If the outcome of the project is only mediocre, but customer expectation is low, then the

outcome of the project may still exceed customer expectation, and, therefore, customer satisfaction can be high. Customer satisfaction is the perceived discrepancy between expectation and realized results.

To understand the effect of perceived realized results to the customer, we analyzed the relationship between budget and time overrun with customer satisfaction. We find that budget overrun is not correlated with customer satisfaction.

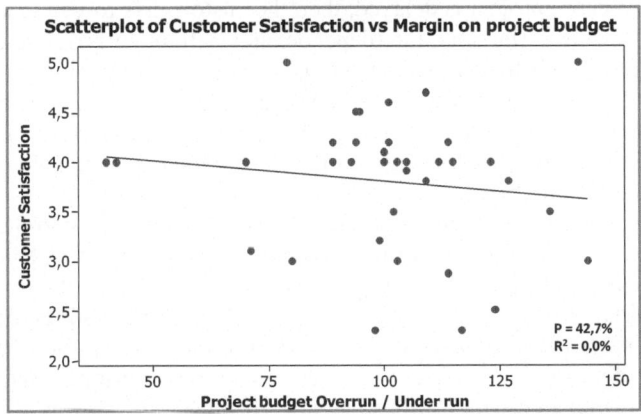

Fig. 3. Customer satisfaction as function of budget overrun

A p-value of 42,7% does not indicate a correlation. This lack of correlation can be explained, when we realize that budget overrun is not necessarily a problem for the customer. In the case of a fixed-price construction, the IT service provider is fully responsible for the budget overrun. In this situation budget overrun may be causing an increase of customer satisfaction, because the customer receives the required functionality, while the overrun costs are paid by the provider. Budget overrun can be correlated with both high or with low customer satisfaction, and is therefore not

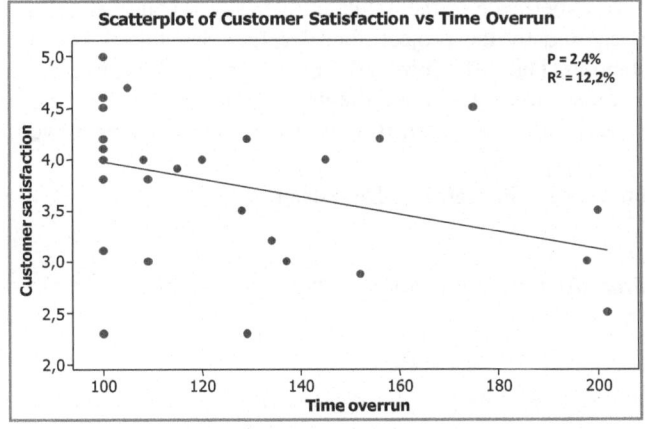

Fig. 4. Customer satisfaction as function of time overrun

related to the perceived value of the project for the customer. See Figure 4 for the analysis of time overrun and customer satisfaction.

We find that this correlation is significant and is described in the following formula:

$$\boxed{?????? ? ???? ? ?????? ? ??? ? ??????n ?} \tag{1}$$

We find that increased time overrun is correlated with decreased customer satisfaction. Contrary to budget overrun, time overrun is always immediately experienced by the customer. When a project is delayed, then the customer is forced to adapt business schedules, the resource planning, interdependencies with other projects, etc.

A further indication that time overrun and customer satisfaction are correlated is also given by the correlations with the project variables 5, 6 and 7 (respectively quality of the project, domain and enterprise architectures) and customer satisfaction, because these three project variables also correlate with the expected value of the project timeframe (success variable 3).

The finding that time overrun and customer satisfaction are correlated, confirms the supposition at the beginning of this paragraph that customer satisfaction is related to the actual outcome of the project. Can we also find correlations that customer satisfaction is correlated to the expectation of the customer? Interestingly, project variable 3 (*Match of certification level of the architect to the level of the project*) does not correlate to budget or time overrun – but correlates with customer satisfaction. Our interpretation is that an architect whose level is matched with the level of the project, manages the expectations of the customer in such a way that it improves customer satisfaction, while less experienced architects does not have this competence.

Therefore, the correlation between customer satisfaction and both time overrun and certification level of the architect (project variable 3) supports our supposition and we can conclude that there are indications that customer satisfaction is influenced by the ability of the architect to manage expectations of the customer and by the time overrun of the project.

Consequences
Customer satisfaction is the result of a comparison of customer expectation and actual outcome of the project. The subjective elements of the customer satisfaction are co-determined by the experience of the architect. If the experience of the architect is lower than the level required by the project, then we find that this is correlated with lower customer satisfaction. The difference is 0,4 point, on a scale from 1 to 5. The objective elements of customer satisfaction are co-represented by the time overrun. The effect is a 0,2 point decrease in customer satisfaction for every 20% additional overrun.

3.8 H_0 Statement VI – Percentage Delivered

H_0 Statement
Usage of solution architecture is not significantly correlated with the expected value of the percentage delivered.

Finding
Project variables 4, 5, 6, 7 and 8 (*Specific experience of the architect, Quality of the project architecture, Quality of the domain architecture, Quality of the enterprise*

architecture and *Quality of the customer's architecture governance process*) test significantly. H_0 statement VI is rejected.

Conclusion

Usage of solution architecture is correlated with a significant increase in percentage delivered. Five of the ten project variables test significantly, which makes percentage delivered one of the success variables that correlates with multiple aspects of the use of architecture.

Broad experience of the architect with the type of engagement correlates significantly with an increase of percentage delivered compared to a situation where the architect has only some experience with the type of engagement. The difference is 8% (92% versus 100%). This project variable explains 11,6 % of the variance in the percentage delivered.

An increase in the quality of the project architecture correlates significantly with an increase of percentage delivered. The difference is between low quality and high-quality project architecture is 12%. This project variable explains 16,9% of total variance of the percentage delivered.

An increase in the quality of domain architecture correlates significantly with an increase of percentage delivered. The difference between low quality and high-quality domain architecture is 13% (92% versus 105%). This project variable explains 13,8% of the total variance of the percentage delivered.

The increase in the quality of enterprise architecture correlates significantly with an increase of percentage delivered. The difference between low quality and high-quality enterprise architecture is 9% (49% versus 103%). This project variable explains 8,6% of the total variance of percentage delivered.

Improved architecture governance correlates significantly with an increase of percentage delivered. The difference between no governance and formal governance is 10% (94% versus 104%). This project variable explains 10,2% of the total variance of percentage delivered.

Significance

P = 1,5% (variable 4)
P = 0,2% (variable 5)
P = 0,6% (variable 6)
P = 2,6% (variable 7)
P = 1,8% (variable 8)

Interpretation

Five of the ten project variables correlate with the success variable percentage delivered. It can well be that the same underlying effect is measured multiple times, but from different angles. For instance, the presence of enterprise architecture and the presence of the domain architecture may be linked by the architectural maturity of the customer's organization. To understand this result more fully, it is necessary to analyze the interaction between project variables (however, see § *4.1 Limitations of the Analysis*). We can conclude that application of enterprise architecture is correlated with a substantial increase in percentage delivered.

Consequences
Analyzing the differences in percentage delivered for the five project variables, we can conclude that usage of solution architecture is correlated with an increase of the percentage delivered of the project with approximately 10%.

3.9 H_0 Statement VII − Functional Fit

H_0 Statement
Usage of solution architecture is not significantly correlated with the expected value of the functional fit.

Finding
None of the variables tests significantly. H_0 Statement VII is not rejected.

Conclusion
The functional fit delivered by projects is not correlated with the use of solution architecture.

Interpretation
This result can be explained by considering the mechanisms of IT project development. It is the business that decides on the functionality of the project; i.e., business answers the *what* question. IT is responsible for building the solution; in other words, IT is responsible for the *how* question. It is therefore explicable that architecture is correlated with the quality of the transformation (as indicated by the other success variables), but not with delivered business functionality.

3.10 H_0 Statement VIII − Technical Fit

H_0 Statement
Usage of solution architecture is not significantly correlated with the expected value of the technical fit.

Finding
Project variable 5 (*Quality of the project architecture*) tests significantly. H_0 statement VIII is rejected.

Conclusion
An increase in the quality of the project architecture correlates significantly with an increase of technical fit.

Significance
P = 0,3%

Interpretation
This result is in line with the interpretation for statement VII. Architecture is correlated with the quality of the transformation, which includes the technical fit (performance, security, availability, etc.).

4 Conclusions and Recommendations

4.1 Limitations of the Analysis

4.1.1 The Role of Second-Order Effects

We found that multiple project variables may correlate with the same success variable. For example, H_0 statement III (Expected value of project timeframe) is correlated with the project variables 5, 6, 7 and 10 (*Quality of the project architecture, Quality of the domain architecture, Quality of the enterprise architecture* and *Architecture compliancy testing*). These variables are correlated with respectively 55%, 44%, 46% and 56% lower time overrun. Can we conclude from these figures that the project variable *Quality of the project architecture* (project variable 5) on its own is responsible for 55% decrease in time overrun? The answer is no, because there are multiple variables or combinations of variables responsible for the decrease in time overrun. See the example below, which illustrates the interaction between project variable 5 and variable 6, for H_0 statement IV -- *Variance of Project Timeframe*.

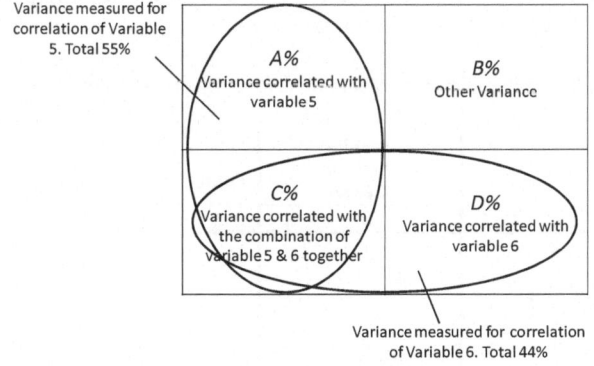

Fig. 5. Example of distribution of variance in project timeframe with two variables

Figure 5 shows a breakdown of the total variance in the success variable project timeframe for two project variables. The total variance is split out into four components: variance that can be explained by the combination of variable 5 and 6, variance that can be explained by variable 5, variance that can be explained by variable 6 and remaining variance that cannot be explained by either 5 or 6. From our measurements we know that A and C together is equal to 55% and that C and D together is equal to 44%, but we cannot determine the value of C, independent from A or D. In other words, we can determine the total variance for with variable 5 or the total variance for variable 6, but we cannot determine the combined effect of both variables, because this is equal to the measured variance of variable 5 (55%) plus the measured variance for variable 6 (44%) minus the combined effect (C%) which is unknown. In reality, we are not dealing with only two variables, but with multiple variables, and the number of second-order interactions between n variables increases quadratic with increasing n. (The number of second-order interactions between n variables is equal to

$n \square h \square \square \square \square \square$.) On top of this, there are third-order interactions, fourth-order interactions, etc.

4.1.2 Measuring Second-Order Effects

The questionnaire used to measure the project variables uses three-level multiple-choice answers. For example, for project variable 5 (the quality of the solution architecture) the possible answers are: high-quality, medium-quality and low-quality. With a survey size of 49 projects, the average sample size for one answer is about 16 (49 / 3) projects. This sample size is important because reliability of an answer depends on this size. To ensure a minimum level of reliability of the answer, a minimum sample size of 6 was chosen. When a sample size is 5 or smaller, then the sample is not used.

If testing a H_0 statement for second order effects, then we need to test it simultaneously for two project variables. In this case, the average sample size becomes 49 / 3^2 = 5,4 projects. The minimum sample size is 6 projects, which means that the average sample size becomes less than the minimum sample size. In addition, variance in sample sizes means that some samples will be even smaller. For instance, a breakdown of the results of project variable 5 (Compliancy Testing) and project variable 10 (Project Architecture) gives the following results[5]:

Table 7. Sample sizes breakdown for project variables 5 and 10

		5. Compliancy Testing Answer			Total
		1	2	3	
10. Project Architecture Answer	1	3	13	8	24
	2	1	3	11	15
	3	0	0	6	6
Total		4	16	25	45

We find that only four of the nine samples sizes are equal or above the minimum sample size of 6. Five of the samples are less than six and cannot be used for testing.

The conclusion from this analysis is that we are not able to measure second order (or higher order) effects. Because of the limited size of the survey (49 projects), any tests for second-order effects are unreliable; the sample size is just too small. To test a H_0 statement simultaneously for two project variables – with the same average sample size of 16 – a survey size of 144 projects is needed.

4.1.3 Consequences

When drawing conclusions for a project variable, then we have to take into account that we are measuring not only a single variable, but we are measuring the effect of this variable combined with the interaction of this variable with other variables.

---

[5] The total number of projects is in this case is 45, because four respondents did not answer this question.

A consequence of this is that we cannot exactly determine which project variables are correlated with an effect. In the example of the previous paragraph, we cannot say that variable 5 is responsible for a 55% decrease of time overrun. We can only say that variable 5, in combination with the variables it interacts with, delivers a combined effect of 55%. However, we do not know the variables it interacts with and we do not know the size of this interaction.

In addition, we are not allowed to combine the results or draw conclusions from the combination of project variables. For example, in the above example we cannot say that a combination of project variable 5 and variable 6, delivers a specific result, or that variable 5 without variable 6 will deliver a different result. The project variables that we have measured are not independent from each other and influence each other in ways we are not able to determine.

However, we must also realize that measuring (only) first-order effects does not imply that these results are not reliable or are not real. The results are real and can be trusted; the limitation of the measurement is that we are not able to determine the exact, individual correlation of project variables with a success variable.

4.2 Results Summary

Usage of solution architecture within software development projects is correlated with the following effects:

Table 8. Overview of main results

(a)	19% decrease in project budget overrun
(b)	Increased predictability of project budget planning, which decreases the percentage of projects with large (> 20%) budget overruns from 38% to 13%
(c)	40% decrease in project time overrun
(d)	Increased customer satisfaction, with 0,5 to 1 point – on a scale of 1 to 5
(e)	10% increase of results delivered
(f)	Increased technical fit of the project results

These results demonstrate that usage of solution architecture is correlated with significant positive effects on software development projects. For instance, result (b) shows that solution architecture is correlated with a reduction of 25% of the percentage of projects with large overrun. This difference is substantial. This degree of improvement justifies the application of development of projects under architecture. With an average project size of € 700.000, this amounts to a saving of approximately € 140.000 for one out of four projects. Typical organizations have a dozen to several hundred IT projects running and on the average project portfolio, this will save millions Euros annually. The same considerations are true for the other results.

Of course, there is a cost associated with building up and maintaining the architecture processes and capability. These costs need to be balanced with the savings. Still, cost is only one of the aspects when taking the choice to implement an architecture function. There are other factors that are also positively influenced by architecture, which are not directly related to financial cost considerations, but are also important for the success of IT within an organization, such as increased customer satisfaction and decreased project time overrun.

It is interesting to note that the identified correlations are all positive: a 'better' value of a project variable correlates with a 'better' outcome of the success variable, for all identified significant correlations. This positive-positive trend gives an intuitive confirmation that the use of architecture is beneficial for projects; use of architecture does not counteract project objectives.

4.3 Other Research

4.3.1 Comparison to Standish Results
The Standish group [19,20]has published a top 10 of project success factors. The 2001 version of the report mentions the following main factors:

Table 9. Overview project success factors (Standish Group [20])

Factor
1.   Executive support
2.   User involvement
3.   Experienced project manager
4.   Clear business objectives
5.   Minimized scope
6.   Standard software infrastructure
7.   Firm basic requirements
8.   Formal methodology
9.   Reliable estimates
10.   Other (Small milestones; Proper planning; Competent staff; Ownership)

Contrary to our findings, this list does not contain any design or architecture factors. An explanation for this could be that at the time of this research (1995-2000), architecture was not widely used or known. The value of architecture was not a topic for IT executives, project managers or project staff and was apparently not identified by the Standish researchers. We feel that architecture should be on this list, because our research shows that architecture is a major project success factor.

Other researchers do value the constructive role of enterprise architecture. For example, the National Research Council states in a review on FBI's Trilogy Information Technology Modernization Program, "if the FBI's IT modernization program is to succeed, the FBI's top leadership [..] must make the creation and communication of a complete enterprise architecture a top priority."[21]. This statement acknowledges the value of enterprise architecture for system development initiatives.

One of the other conclusions from the original Standish Chaos report [19] is that the success rate and the size of the project are linked. The lower the project cost, the higher the success rate. They provide the following figures:

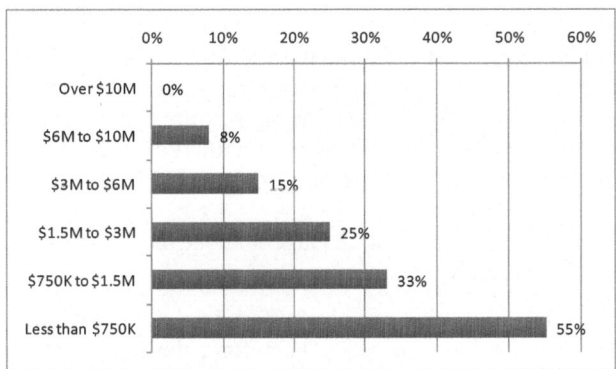

Fig. 6. Project success rates (Standish Chaos Report [19])

A correlation of the size of the project with budget overrun provides the following result:

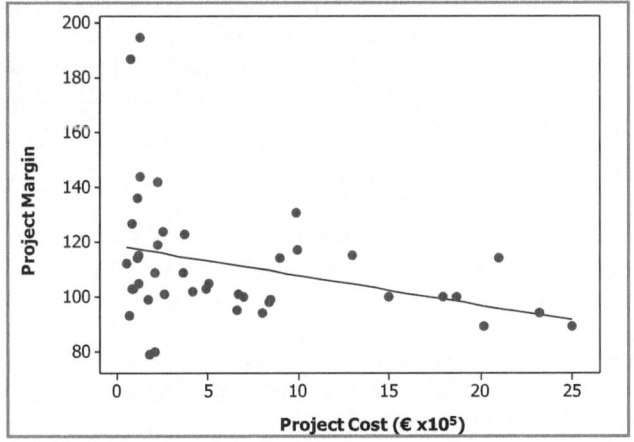

Fig. 7. Correlation between project cost and margin

The vertical axis describes the percentagewise budget overrun or under run. The horizontal axis describes the total cost of the IT project (x 10^5) in Euro's.

Table 10. Key figures for the linear correlation between project cost and margin

Figure	Value
p	0,035
r^2	8,4%
Correlation formula	*Project Margin = 118,6 – 1,1.10^{-5} * Project Cost*

The correlation that we find between project cost and project margin is that projects become *more* successful with increasing project size. This is a contradiction to the findings of the Standish report, because they find their projects become *less* successful with increasing project size. The figures are not exactly comparable, because Standish defines project success as a combination of on time, on budget and with sufficient functionality. Our correlation only considers cost overrun. Still, the trend is clearly contradictory.

In discussion with project and risk managers about the reason for our finding, the following explanations are given:

1. For small projects, the initial planning effort in determining the project cost is much smaller than for large projects. Therefore, the project cost estimations for small projects are less reliable and the complexity of the project may be underestimated.
2. For small projects, it is very difficult to overcome a project setback within the existing budget. If a small, two-month project has a setback which delays the project for one month, than the budget overrun in absolute terms may be small, but percentagewise the overrun is 50%. For large projects, this type of small setbacks can be absorbed within the existing project budget and the risk margins.
3. For large projects, you have the time to rethink (part of) the solution and learn from lessons earlier in the project. For small projects, if you are halfway through the project and then find out that the original solution needs adjustment, there is no time or budget to redesign.

These arguments provide an explanation for finding that increasing project size correlates with higher project success.

When trying to explain the discrepancy between our results and the findings from the Standish report, then we must realize that our maximum project size is € 2,5M, while the Standish report examines projects with maximum size over * 10M. This variation in maximum size may explain the difference. The arguments above describe the reasons why small projects (€ 1M) have high overruns; these arguments are not relevant for projects above € 2,5M. It is possible that the decreasing trend reverts to an increasing trend for larger projects.

4.4 Applicability and Conclusions

4.4.1 Applicability of the Results

The study that we conducted was carried out in a rather uniform context. Individual projects are relatively small (less than € 2,5M) and all projects were executed in the context of a commercial IT service provider. Furthermore, the number of projects within the scope of the study is a rather limited. This type of research benefits from a large survey base. Several hundred projects would be better; several thousand projects would still be better. The question is then whether the type of effects and the direction of the effects that we have measured (architecture lowers budget overrun, lowers time overrun, increases the percentage delivered, etc.) are valid in the general situation. Based on this one study, we cannot provide definitive statements on this.

Nevertheless, one of the major findings of the analysis of § 4.2 is that: "a 'better' value of a project variable correlates with a 'better' outcome of the success variable, for all identified significant correlations. This positive-positive trend gives an intuitive confirmation that the use of architecture is beneficial for projects; use of architecture does not counteract project objectives." None of the identified correlations between architecture project variables and success variables counteracts project success; all correlations are in the same positive direction.

4.4.2 Overall Conclusion
We did not identify correlations between project variables and success variables with a negative effect on the success variable. To the contrary, we identified multiple positive correlations. We therefore conclude that the use of solution architecture is correlated with a substantial improvement of several key success variables. Based on this finding, our main conclusion is that we can confirm our initial hypothesis that for custom software development projects with a maximum cost of € 2,5M, the use of solution architecture is correlated with improved project results.

Acknowledgements

A survey like this cannot be conducted without the cooperation of many people. We would like to thank everyone who contributed to this research. Our gratitude goes especially to the following persons: André Weber, Bernard Hüdepol, Edgar Giessen, Edwin Kok, Frank Harmsen, Ger Donners, Goossen Foppen, Hans Schevers, Jack van Eijk, Jan Joosen, Jos Smit, Jouke Hopperus Buma, Kaeso de Jager, Maickel Sweekhorst, Marco Folpmers, Mark Grimberg, Mathieu Hagen, Max Stahlecker, Nico Toussaint, Ron Tolido, Rudolf Jurgens, Tom Vanderwiele, Ton Hardeman, Tonny Wildvank, Victor van Swede, Wil van Hamersveld, Wouter Schmitz and the project managers who took the time to fill in the questionnaire.

References

1. TOGAF: The Open Group Architecture Framework, Version 8.1 Enterprise Edition. TOGAF (2004)
2. Lankhorst, et al.: Marc: Enterprise Architecture at Work, Modeling, Communication and Analysis. Telematica Instituut, Enschede (2005)
3. Rijsenbrij, D., Goedvolk, H.: Integrated Architecture Framework. White Paper (1999)
4. Parker, M., Trainor, E., Benson, R.: Information Strategy and Economics. Prentice-Hall, Englewood Cliffs (1988)
5. Oirsouw, R., Spanderman, J., Vries, H.: Informatie Economie, Investeringsstrategie voor de Informatievoorziening (1993)
6. Kerzner, H.: Project Management: A Systems Approach to Planning, Scheduling, and Controlling. Wiley, Chichester (2003)
7. Office of Government Commerce: PRINCE2. Office of Government Commerce (accessed 2008), http://www.ogc.gov.uk/methods_prince_2.asp
8. Barnes, J.: Implementing the IBM Rational Unified Process and Solutions: A Guide to Improving Your Software Development Capability and Maturity. IBM Press (2007)
9. DSDM: Enabling Business Agility. In: DSDM Consortium, http://www.dsdm.org/10 (accessed, August 2008)

10. SEI: CMMI for Development Version 1.2. Carnegie Mellon University - Software Engineering Institute (2006)
11. DSDM and TOGAF: DSDM and TOGAF Joint White Paper (2003)
12. Ambler, S.: Enterprise Unified Process (EUP) (accessed, January 2009)
13. Piselo, T., Strassman, P.: IT Value Chain Management - Maximising the ROI from IT investments. Standish Report (2003)
14. Wagter, R., Berg, M., Luijpers, J., Steenbergen, M.: Dynamic Enterprise Architecture: How to Make It Work. Wiley, Chichester (2005)
15. Wohlin, C., Andrews, A.: Prioritizing and Assessing Software Project Success Factors and Project Characteristics using Subjective Data. Empirical Software Engineering: An International Journal 8(3), 285–308 (2003)
16. Jurgens, R.: The Added Value of Enterprise Architecture, M Sc Thesis., University Delft (2008)
17. Wideman, M.: Improving PM: Linking Success Criteria to Project Type, http://www.maxwideman.com (accessed July 02, 2008), http://www.maxwideman.com/papers/improvingpm/dimensions.htm
18. Pyzdek, T.: The Six Sigma Handbook, Revised and Expanded. McGraw-Hill, New York (2003)
19. Standish Group International: CHAOS: A recipe for success (1999)
20. Standish Group International: Extreme Chaos (2001)
21. McGroddy, J., Lin, H. (eds.): A Review of the FBI's Trilogy Information Technology Modernization Program, Washington DC (2004)
22. Bass, L., Clements, P., Kazman, R.: Software Architecture in Practice. Addison-Wesley, Reading (2003)
23. IEEE: The IEEE 1471-2000 standard - Architecture Views and Viewpoints (2000)
24. Maes, R., Rijsenbrij, D., Truijens, O.: Reconsidering Information Management Through a Generic Framework, Amsterdam (1999)
25. Martin, J., Leben, J.: Strategic Information Planning Methodologies. Prentice-Hall, Englewood Cliffs (1989)
26. Capgemini: Architecture and the Integrated Architecture Framework, http://www.capgemini.com (accessed, December 2007), http://www.capgemini.cm/resources/thought_leadership/architecture_and_the_integrated_architedcture_framework
27. Kruskal, W., Wallis, W.: Use of ranks in one-criterion variance analysis. Journal of the Americal Statistical Society(45) (1952)
28. Levene, H.: In Contributions to Probability and Statistics: Essays in Honor of Harold Hotelling. Stanford University Press (1960)
29. Lindman, H.: Analysis of variance in complex experimental designs. W. H. Freeman & Co., San Francisco (1974)
30. Zachman, J.: A Framework for Information Architecture. IBM Systems Journal (1987)
31. Montgomery, D.: Introduction to Statistical Quality Control. John Wiley and Sons, Chichester (2005)
32. Moore, M., Kazman, R., Klein, M., Asundi, J.: Quantifying the Value of Architecture Design Decisions: Lessons from the Field. Software Engineering (2003)

Quality Enhancement in Creating Enterprise Architecture: Relevance of Academic Models in Practice

Agnes Nakakawa[1], Patrick van Bommel[1], and H.A. Erik Proper[1,2]

[1] Institute of Computing and Information Sciences, Radboud University Nijmegen
Heyendaalseweg 135, 6525 AJ Nijmegen, The Netherlands
`A.Nakakawa@science.ru.nl, pvb@cs.ru.nl`
[2] Capgemini, Papendorpseweg 100, 3500 GN Utrecht, The Netherlands
`e.proper@acm.org`

Abstract. This chapter presents an explicit approach, that is both theory and practice driven, to support evaluation and collaboration activities when creating enterprise architecture. The approach will be applicable in addressing evaluation and collaboration related aspects in two primary phases of the Architecture Development Method (ADM) of The Open Group Architecture Framework (TOGAF). The phases of interest are preliminary phase (defining architecture principles) and phase A (creating architecture vision). These two phases involve activities where evaluation of alternatives and collaboration among key stakeholders and enterprise architects, are paramount. Based on theoretical insights, a collaboration process to facilitate the steps in the formulated approach has been developed. Both the approach and the process design for its realisation, have been evaluated by exposing them to practitioners. This was done using structured walkthoughs. Insights from these walkthrough sessions with experienced enterprise architects, were used to enrich the theoretical models. Generally this chapter aims at demonstrating how theoretical models, enriched with experiences from industry, can fill the currently existing lack of profound analysis of success factors for enterprise architecting. Note that this lack exists in both academia and industry.

Keywords: Enterprise Architecture, Design Alternatives, TOGAF, Collaboration Engineering, Practical Relevance.

1 Introduction

While making decisions regarding an enterprise transformation, stakeholders desire to understand the impact of the transformation on their concerns and the risks associated with current and future strategies of the enterprise [23]. Any changes in an organisation's strategy and business goals considerably affect all domains of the enterprise [15], and its corresponding partnerships or collaborations. An example of a rewarding enterprise transformation is enterprise

E. Proper, F. Harmsen, and J.L.G. Dietz (Eds.): PRET 2009, LNBIP 28, pp. 109–133, 2009.

architecture development. While the debate on the definition of (enterprise) architecture continues, discussions in this chapter concentrate on the definition provided by The Open Group Architecture Framework (TOGAF). This is because TOGAF is freely available, neutral towards tools and technologies, and is a detailed approach for supporting architecture development [32]. Architecture is *"(1) a formal description of a system, or detailed plan of the system at component level to guide its implementation; (2) the structure of components, their inter-relationships, and the principles and guidelines governing their design and evolution over time"* [32].

Since business essentials are more stable than specific solutions that are found (or sought) to address current (or emerging) problems, architecture assists in guarding business essentials while permitting maximum flexibility and adaptability [15]. Moreover, objects (such as an enterprise) designed under architecture offer improved performance regarding adaptability, integration, understandability, and agility among others [37]. The internal drive of an organisation to adopt enterprise architecture practice, is to effectively execute its strategy and optimise its operations [15]. However, this can be sufficiently achieved if, when creating enterprise architecture, possible design alternatives are generated, evaluated, and *appropriate* as well as *efficient* ones, are selected. Appropriate in this context refers to the suitability of the architecture to address its planned purpose and realise organisation objectives. Whereas efficiency is the ability of the architecture results to address stakeholders' concerns [23].

The endeavor of evaluating design alternatives will further yield better results if it is done in a collaborative context, involving enterprise architects and all organisation key stakeholders. In this chapter we hereby explore the practical relevance of formulating a two-fold approach that we refer to as Collaborative Evaluation of Enterprise Architecture Design Alternatives (CEEADA). The approach is two-fold in the sense that it addresses both collaboration and evaluation related aspects when creating enterprise architecture. CEEADA is a theory based approach that has recently been enriched by practice driven insights from practitioners. In this chapter we explain in detail the theoretical underpinnings of CEEADA, and discuss how insights from experienced enterprise architects were used to enrich CEEADA. These practice based insights were obtained through conducting structured walkthrough sessions with enterprise architects.

The chapter hence fills the gap, in both academia and industrial practice, of two significant needs in enterprise architecture development. First is the need for ensuring collaboration between architects and key stakeholders during enterprise architecture development. This need has been emphasized by several researchers and practitioners (e.g. in [1,2,14,21,23,26,27,34]), but a sustainable, explicit, and consistent approach for sufficiently addressing this cause is absent in both academia and practice. Second is the need for evaluating enterprise architecture design alternatives and performing trade-off analysis when creating enterprise architecture. This need has also been emphasized by researchers and practitioners (e.g. in [23,32]), but an explicit and consistent approach for sufficiently addressing this cause is absent as well, in both academia and practice.

The remainder of this chapter is organised as follows. Section 2 discusses efforts by researchers and practitioners towards evaluation of artifacts in the domain of enterprise architecture. Section 3 presents theoretical underpinnings of CEEADA. Section 4 presents how Collaboration Engineering was used to design a collaboration process that can enable organisations to realise CEEADA in a sustainable way. Section 5 discusses the applicability of the approach within TOGAF's Architecture Development Method (ADM). Section 6 presents practice driven insights from enterprise architects into the approach, and illustrates modified CEEADA models. Section 7 concludes the chapter.

2   Evaluation Efforts in Enterprise Architecture Domain

This section discusses existing work on quality of artifacts in enterprise architecture practice. It also highlights aspects regarding quality achievement in the architecture creation process, that have been given insufficient attention.

A good (or high quality) enterprise architecture offers insights into balancing business requirements and transforming enterprise strategy into daily operations [15]. However, there are several interpretations of the correctness (in this context appropriateness) of an architecture [24]. The acceptability and appropriateness of an enterprise architecture vary across organisations, since they are relative to business requirements and stakeholders' concerns. Actually the kind of results expected from the architecture effort depends on the purpose of the architecture [23].

Existing work on evaluation of artifacts in enterprise architecture domain has mainly concentrated on measuring quality and benefits or return on investment of enterprise architecture. For example, in [28] a framework is presented, based on balanced scorecard approach, for enabling corporate management to identify and measure benefits of enterprise architecture. In [31], quantitative benefits of architecture are explored, and it is demonstrated how architecture may substantially reduce project risks and corresponding costs. In [33] an instrument is presented, based on Sogeti's DYnamic Architecture method, for measuring the quality of the process for enterprise architecture development. Moreover, in [35] an instrument is presented, based on DYnamic Architecture method, for determining the quality of (tangible) products delivered by enterprise architects.

A formal approach for verifying and validating the relevance and suitability of a developed enterprise model is also presented in [6]. However, since enterprise architecture addresses company-wide integration [20], evaluation and validation of its model(s) could be complex especially if, when creating these models, insufficient quality assessment was done on its individual (tangible and intangible) components. Therefore, although it is significant to do a quality check on enterprise architecture products before they are deployed [35], evaluation of possible design alternatives during the creation of these products is equally significant. Actually in the context of TOGAF, it is recommended that there should be frequent validation of results for the entire ADM cycle, and for a particular completed phase of the ADM [32]. Enterprise architecture benefits can better be reaped if, when creating architecture, the quality of decisions behind its components are also put into consideration. Such a reflection has been given little attention so far.

Additionally, it is reported that the quality of enterprise architecture products can be improved based on expectations of organisation stakeholders [34]. Such expectations can be obtained and comprehended through effective collaboration between enterprise architects and stakeholders during the architecting process. Literature hardly reveals efforts towards how these aspects can be handled to improve the quality of the process for creating enterprise architecture. Therefore, our research scheme generally focuses on achieving a method that can be used within an enterprise architecture framework (particularly TOGAF ADM), to address collaboration related aspects and evaluation of design alternatives, when creating enterprise architecture. Such a method will be significant towards filling the gap, which is reported in [23], of the lack of scientific research on success factors for enterprise architecting.

3   CEEADA in Creating Enterprise Architecture

This section presents theoretical underpinnings of CEEADA, quality related variables in the process of creating enterprise architecture, and an explicit approach for balancing such variables in order to realise CEEADA.

Creating enterprise architecture generally involves understanding the purpose of the architecture effort, determining deliverables, monitoring planned architecture context, creating shared conceptualisation among stakeholders, designing the architecture creation process, determining impacts, and communicating the architecture [23]. Several enterprise architecture frameworks are in place to guide the architecture creation process. Yet some enterprise architecture projects may fail to deliver as planned, due to a number of challenges.

Challenges that enterprise architects and organisations face during enterprise architecture development originate from political, project management, and organisational problems and weaknesses, rather than technical aspects [16]. Such challenges can be steadily addressed by gradually building consensus among stakeholders through effective collaboration, and encouraging informed evaluation of possible design alternatives when creating enterprise architecture. These aspects are significant during the high level definition of the architecture. This is because if they are not intensively addressed at that point, it will negatively affect the quality of any intended evaluation of alternatives and collaborative work in the subsequent architecture activities. However, as discussed in section 2, literature hardly reveals an explicit and consistent approach for addressing these two aspects in the enterprise architecture domain.

We therefore offer theoretical insights (guided by design science) into improving the process of creating enterprise. In this paragraph, we briefly describe design science based on [11,12,13,30]. Design science is a paradigm for problem-solving that was pioneered by Simon in 1969. It is concerned with the creation and evaluation of IT artifacts (i.e. constructs, models, methods, and instantiations) for solving identified organisational problems. It also enables formulation of new artifacts that offer opportunities for improving practice prior to practitioners recognising any problem with the existent way of working. Creation of these artifacts is supported by pre-existing theories, frameworks, instruments, constructs, models, methods, and instantiations.

Thus, in devising an approach for CEEADA, we first draw upon the causality analysis theory to perform a cause-effect analysis of key variables for improving the architecture creation process. This is because explaining an event usually involves explaining its cause, and an analysis of the relation between cause and effect of events is essential to several formations of theory (i.e. conjectures, models, frameworks, or body of knowledge) [10]. Causality analysis will thus help in the formulation of models to realise CEEADA.

3.1   Cause-Effect Review in Creating Enterprise Architecture

From [1,2,14,19,21,23,26,27,32,34], we identify variables that are key to quality enhancement in creating enterprise architecture. As figure 1 shows, these variables include: the quality (appropriateness and efficiency) of an enterprise architecture, the quality (appropriateness and efficiency) of an enterprise architecture component, the quality of the evaluation process of architecture design alternatives, the quality of collaboration among key stakeholders, the quality of enterprise architecture creation process, the level of consensus on evaluation

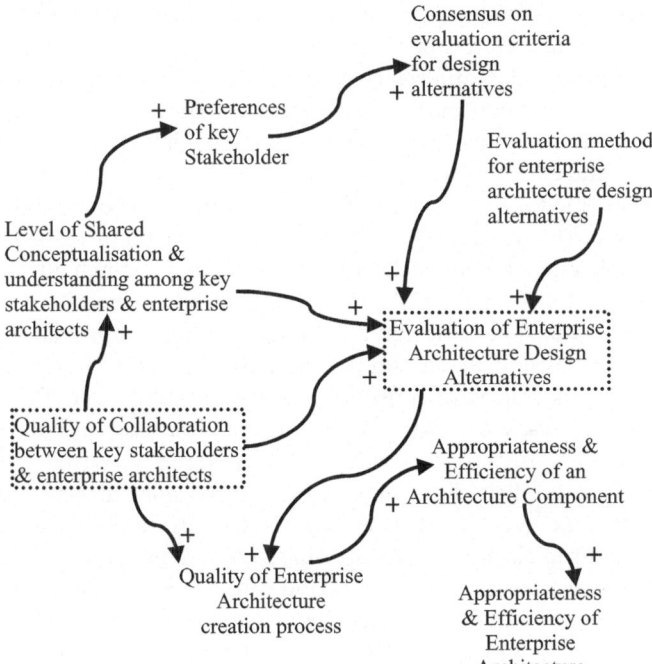

Fig. 1. Cause-Effect Analysis in Creating Enterprise Architecture

criteria for enterprise architecture design alternatives, the evaluation method for enterprise architecture design alternatives, and the level of shared conceptualisation and understanding of organisation problem and solution aspects among key stakeholders.

In the following explanations for figure 1, we concur with Gregor that *various arguments for causality are not mutually exclusive and at di erent times and in different circumstances we will rely on di erent reasons for ascribing causality* [10].

The quality of the process of creating enterprise architecture can be improved by evaluating enterprise architecture design alternatives, and by encouraging effective collaboration among key stakeholders and enterprise architects. The reason for evaluating (design) alternatives is to search for optimal or satisfactory solutions [29,30]. Such solutions can be viewed as high level solutions or low level unit components of the high level solution. In this context, architecture components include principles, models, and views [23]. In our view, there are design alternatives regarding each of these components during the architecture creation process. Therefore, evaluating them and selecting satisfactory and optimal ones, will add value to the architecture creation process.

Better still, evaluating such alternatives in a collaborative context leads to better decisions. This is because successful problem solving and decision making in organisations often requires joint expertise [19]. Moreover, maximum effectiveness of the architecture function is only attainable if stakeholders efficiently collaborate towards a shared goal [34]. Therefore, effective collaboration adds value to the process of evaluating enterprise architecture design alternatives.

In [31], it is demonstrated how the quality of (enterpise) architecture is one of the key inputs for high customer satisfaction in a given project. Logically, if the quality of architecture affects customer satisfaction, then the quality of the process of creating architecture indirectly affects customer satisfaction. Our definitions of appropriateness and efficiency of enterprise architecture (see section 1), are closely related to customer satisfaction. Therefore, as shown in the lowest part of figure 1, an improvement in the quality of the architecture creation process leads to selection of appropriate and efficient architecture components, which ultimately results into creation of an appropriate and efficient enterprise architecture.

Additionally, evaluation of design alternatives can be improved by: (1) a high level of shared conceptualisation and understanding of enterprise aspects among stakeholders, (2) a high level of consensus on evaluation criteria for design alternatives, and (3) the evaluation method for design alternatives. Full commitment of stakeholders in an initiative is often guaranteed if a shared goal has been acquired [19]. This implies that achieving a shared goal directly improves the priorities of stakeholders. This in turn results into an increased level of consensus on evaluation criteria for design alternatives. For example, results obtained after ranking of alternatives some evaluation criteria, are often consistent with a stakeholder's objectives and preferences [9]. Ultimately, the evaluation of design alternatives is directly and indirectly improved by an increased level of shared conceptualisation and understanding of aspects among stakeholders.

Furthermore, the level of shared conceptualisation and understanding can be increased by effective and efficient collaboration between stakeholders and enterprise architects. This is because collaboration is a joint effort of stakeholders towards achieving a goal, and the probability of acquiring shared and supported goals is higher when stakeholders make this joint effort [19]. On the other hand, mutual understanding is a requirement for architects and stakeholders to improve their collaboration and make the architecture function effective [34]. This hence reveals a recursive relation between shared understanding and collaboration.

The causal relations explained above cannot be sufficiently measured in isolation, but a hypothesis can be drawn, and a synthesis formulated from such relations, such that they are measured in an integrated and meaningful manner. This is possible because the knowledge of causal relations enables predictions to be made from theory [10]. Therefore from figure 1, and the underlying reasons for its factors, the following predictions are made with the focus of improving the quality of the process of creating enterprise architecture.

Since key stakeholders have diverse concerns and views, they could first acquire a shared conceptualisation and understanding of enterprise aspects. A shared conceptualisation and understanding is a basis for evolution of an enterprise [23]. A shared understanding will consequently guide the determination of common and explicit criteria for evaluating enterprise architecture design alternatives, the identification and validation of possible design alternatives, the evaluation of such alternatives, and the selection of appropriate and efficient ones. This approach for enabling CEEADA is illustrated in figure 2, decomposed and characterised in figure 3, and explained thereafter.

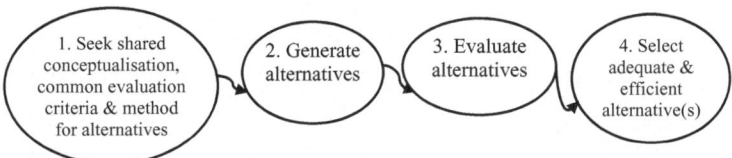

Fig. 2. Collaborative Evaluation of Enterprise Architecture Design Alternatives

In the middle part of figure 3 we see the pattern for CEEADA consisting of four steps shown by dashed boxes. Above the dashed boxes we see the decomposition of tasks for the four steps, and sub activities involved in each step are shown. Below the dashed boxes we show the characterisation of CEEADA according to Simon's generic decision making process. The pattern for CEEADA has its roots in the generic decision making paradigm introduced by Simon in 1960 in [29]. Simon structured all decision making tasks to comprise of three phases, i.e. intelligence, design, and choice. Intelligence is concerned with investigating an environment for circumstances that call for decision or intervention. Design is concerned with devising possible courses of action or possible decision alternatives to solve the problem or to improve the environment. Choice is concerned with choosing a particular course of action or decision alternative from

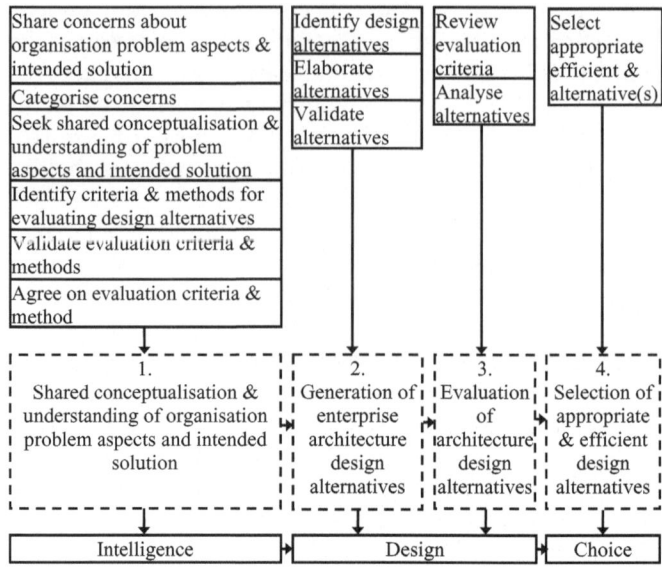

Fig. 3. CEEADA Pattern Decomposition and Characterisation

those available. As figure 3 shows, step 1 of the pattern for CEEADA is characterised as Simon's intelligence phase, steps 2 and 3 are characterised as design,
and step 4 is characterised as choice. The following sections describe these steps
in detail.

3.2 Shared Conceptualisation◇Common Evaluation Criteria

Agility as a key requirement in several business lines is often hindered by organisation stakeholders being uninformed about their own products, services, and
capabilities; and lacking a common understanding and governance of data resources [23]. Stakeholders should understand aspects related to data and control
flow, as well as decisions that will affect the organisation's overall performance
[16]. Although several companies still lack an integrated view of their enterprise,
the architecture process helps to raise stakeholders' awareness of business objectives and information flow [15]. However, stakeholders' awareness of these key
aspects, during the architecting effort, is not an automatic achievement.

Thus, the architecting process should be 'open' in the sense that participation of stakeholders is encouraged [1,2]. This openness calls for collaboration
between architects and organisation stakeholders. Moreover, although collaboration between architects and stakeholders is problematic, it can be effective if
also architects acquire a good understanding of the goals of the stakeholders
[34]. Figure 1 shows that effective collaboration between stakeholders and architects during enterprise architecting enhances a shared conceptualisation and
understanding of all key aspects.

The enterprise architecting process requires all involved actors to speak a common and identical technical language, and to have a shared understanding of what the architecture is supposed to do [1]. Shared understanding involves: sharing knowledge, sharing meaning about the knowledge, mutual learning (people learning from each other to advance their knowledge and the group knowledge), and understanding of mutual differences or conflicts [17]. It is essential for stakeholders to acquire a shared conceptualisation and understanding about 'the *as-is* situation'; 'the *to-be* situation'; and any constraints that should be met by the architecture [23]. Open modeling, sharing models, and frequent communication with stakeholders can enable the architect to steadily eliminate the different implicit views that individual stakeholders have regarding the intended system [21].

Additionally, literature hardly reveals explicit criteria for evaluating enterprise architecture design alternatives during the architecting process. Evaluation criteria for design alternatives often vary across organisations depending on the organisation's mission and vision. This therefore calls for stakeholders and enterprise architects to identify, evaluate, and agree on explicit criteria and a method for evaluating enterprise architecture design alternatives. This is possible if a shared conceptualisation and understanding of organisational problem aspects has been attained.

3.3   Generation of Design Alternatives

Designing a system (in this context, enterprise architecture) consists of determining its requirements and devising feasible specifications that satisfy the agreed on requirements [37]. In the endeavor to optimally fulfill these requirements and specifications, design alternatives arise. Enterprise architecture comprises of four major types of architectures, i.e., business, data, applications and technology [32]. Logically enterprise architecture design alternatives arise from these architectures types, and from the phase of defining framework and principles (TOGAF's preliminary phase), and creating architecture vision (TOGAF's phase A). Section 5 expounds this. Enterprise architecture design alternatives can be generated at different phases of architecture development, depending on the enterprise architecture framework that has been adapted.

We give two reasons for collaborating with key stakeholders even at this step. First, is the creativity that collaboration offers during problem solving [7]. Creativity is a key input to generating design alternatives of a solution. Logically generation of design alternatives can be more fruitful if key stakeholders and enterprise architects have acquired a shared conceptualisation and understanding of problem and solution aspects. Second, involvement of key stakeholders at this step gradually builds commitment and consensus among them. This is because during the intelligence and design phases of decision making, commitment of actors to a new course of action can gradually evolve [29].

Generation of alternatives involves identifying, elaborating, and validating possible architecture design alternatives. Elaboration of design alternatives involves adding relevant detail to an alternative, preparing it to be evaluated.

Vague concepts in an enterprise architecture should be translated to a detailed level such that the architecture is understandable and agreed on by all stakeholders [14]. Detail does not need to be restricted to only the overall enterprise architecture but can be carried over to its constituent components and their respective design alternatives. Detailed alternatives enable informed evaluation of alternatives to be performed.

Validation of design alternatives involves investigating an alternative for its feasibility. Validation of alternatives is most likely to be affected by the information available for each design alternative. The lack of knowledge and misunderstanding of particular features and information from a system (say an enterprise) or its environment consequently limits the verification and validation of (enterprise) model(s) [6]. This further explains why stakeholders and enterprise architects should effectively collaborate in the generation of design alternatives, and above all, have a shared conceptualisation and understanding of enterprise aspects.

3.4   Evaluation of Design Alternatives

Evaluation involves assessing the appropriateness and efficiency of each validated design alternative, with respect to predefined common evaluation criteria, using a common evaluation method. Often the predefined evaluation criteria may require revision, hence the need for consensus on any amendments. In decision making some decisions may be too complex for an individual to understand all implications [19] regarding each decision alternative. Hence the need for collaboration among enterprise architects and stakeholders during the evaluation of design alternatives. Stakeholders' involvement in the evaluation of design alternatives gradually increases consensus among them.

Before evaluating design alternatives, the type of evaluation problem must be understood because it determines the evaluation methods to be used. According to [8], evaluation problems are categorised into three: (1) Choice problems, involve *selecting of a subset of actions, as small as possible, in such a way that a single action may be nally chosen* , (2) Ranking problems, involve *ranking of all the actions belonging to a given set of actions from the best to the worst* , (3) Sorting problems, involve first defining a set of categories depending on some typical features, and then *assigning each action to one of the pre-de ned categories* .

From these problem types, the idea of collaboratively evaluating enterprise architecture design alternatives is a "Sorting-Ranking-Choice" problem. This is because in order to realise CEEADA, at least one of the three problems must be encountered at different instances when creating enterprise architecture. For example when defining architecture principles, a ranking problem could be encountered; yet when defining architecture vision, both sorting and choice problems could be encountered. Therefore, a "Sorting-Ranking-Choice" problem in CEEADA would generally appear as follows.

1. In a sorting problem context, categories of design alternatives at a given phase of architecture development would be defined. For example using TO-GAF ADM, categories of design alternatives at phase A (architecture vision) would include aspects regarding scope, constraints, baseline architecture, and target architecture. Then each action (in this case design decision alternative), would be assigned to a category where it can be further assessed.
2. In a ranking problem context, all possible enterprise architecture design alternatives are ranked from best to worst. Where ranks are based on stakeholders' priorities and quality value judgements.
3. In a choice problem context, a subset of architecture design alternatives can be selected, based on stakeholders' value judgements and priorities, from which a single alternative will be finally chosen.

3.5   Selection of Appropriate and Efficient Design Alternatives

The focus at this step is to select design alternatives that will collectively result in optimal business operations and an appropriate and efficient enterprise architecture. Although it is difficult to satisfy all stakeholders [34], a solution embraceable by key stakeholders can be sought.

Two situations may arise at this step, depending on the phase of architecture development and the type of evaluation problem encountered in that phase. (1) Only one alternative may be required, for example the alternative with the highest score or rank, making the selection step to be trivial; or (2) more than one alternative may be required. In case 2, the remaining alternatives may be assessed using additional evaluation criteria.

4   Collaboration Engineering

Literature [4,25] reveals sustainable approaches (i.e. collaboration engineering and group model building scripts) that can be used to enable execution of steps in CEEADA. This section therefore presents an attempt of applying collaboration engineering to this cause.

Collaboration engineering is an approach used for designing re-usable collaboration processes that yield predictable success for recurring mission-critical tasks, and the deployment of such processes for execution by practitioners rather than skilled facilitators [5,18,36]. Relevant facilitation skills, knowledge of group support systems, and group dynamics can be transferred to practitioners using this approach, since skilled facilitators are an additional cost to organisations [4,18]. In a collaboration process, participants undergo a reasoning process that comprises of a series of activities referred to as basic patterns of collaboration or thinking [4]. Six general patterns of collaboration are defined in [5] as follows.

1. *Generate*, moving from having fewer concepts to more concepts as shared by the group.
2. *Reduce*, moving from having many concepts to a focus on fewer concepts that the group considers worthy of further attention.

3. *Clarify*, moving from having less to more shared understanding of concepts and phrases used to express them.
4. *Organise*, moving from less to more understanding of the relationships among concepts the group is considering.
5. *Evaluate*, moving from less to more understanding of the relative value of the concepts under consideration.
6. *Build consensus*, moving from having fewer to more group members willing to commit to a proposal.

Each pattern of collaboration is created by a unit known as a ThinkLet, which defines the group support system to use; how to configure it; and a clear sequence of events and instructions for the group to follow [4]. Therefore, thinkLets are building blocks for designing collaboration processes [17,18].

To formulate a collaboration process for CEEADA, the following design approach as described in [17,36] was used.

1. *Task diagnosis*, determining the goal and deliverables of a collaboration process.
2. *Task decomposition*, determining the basic activities for realising the process goal.
3. *ThinkLet choice*, matching each basic activity with a thinkLet using some criteria.
4. *Agenda building*, preparing all relevant information for validating the process and graphically representing it in a Facilitation Process Model (FPM). The FPM shows *"the logic of the ow of the collaboration process from activity to activity"* [17].
5. *Design validation and evaluation*, using walkthroughs, pilot testing, simulation, and expert evaluation.
6. *Documentation*.

Under task diagnosis, the goal of our collaboration process is to realise CEEADA when creating enterprise architecture. Our results for task decomposition, thinkLet choice, and agenda building, in CEEADA, are summarised in table 1. The FPM for CEEADA is illustrated in figure 4. The building patterns used in table 1 and fig. 4 are described in [36]. Initial versions of table 1 and figure 4 are presented in [22].

5   Relevance of CEEADA in Practice

This section discusses how quality of output from the first two phases of TOGAF ADM can be improved by applying CEEADA. In sections 5.1, 5.2, and 5.3, a brief report is first given on the steps involved in each phase, as presented in [3,32], then the applicability of our approach in that particular phase is discussed.

5.1   Defining Framework and Principles

This TOGAF phase generally involves: (1) defining the framework to be used (i.e. adapting the ADM); (2) reviewing (pre-existing) business principles, goals,

Table 1. Key Activities, Patterns of Collaboration, and ThinkLets

#	Activity Description	Deliverable	Pattern of Collaboration	ThinkLet
0	Prepare for architecture development sessions	Architecture Development information & sensitization	-	-
SESSION ONE – Shared Conceptualisation & Common Evaluation Criteria				
1A	Introduction/Briefing	Guiding information	-	-
1B	Share concerns	Concerns	Generate	LeafHopper
1C	Categorize concerns	Categories of concerns	Reduce & Clarify	FastFocus
1D	Discuss concerns while seeking shared conceptualization & understanding of enterprise aspects	Shared understanding of aspects & a common view of the enterprise	Build Consensus	CrowBar
1E	Identify criteria & methods for evaluating design alternatives	Evaluation criteria & methods	Generate	Free Brainstorm
1F	Categorize criteria & methods	Categories of criteria & methods	Reduce & Clarify	FastFocus
1G	Evaluate criteria & methods	Evaluated criteria & methods	Evaluate	StrawPoll
1H	Agree on evaluation criteria & method	Common evaluation criteria & evaluation method	Build Consensus	MoodRing
SESSION TWO – Generation of Enterprise Architecture Design Alternatives				
2A	Identify design alternatives	Design alternatives	Generate	Comparative Brainstorm
2B	Elaborate alternatives	Elaborated alternatives	Generate	TheLobbyist
2C	Validate alternatives	Validated alternatives	Evaluate	StrawPoll
SESSION THREE – Evaluation and Selection of Design Alternatives				
3A	Evaluate alternatives	Evaluated alternatives	Evaluate	MultiCriteria
4A	Select appropriate & efficient alternative(s)	Appropriate & efficient design	Build Consensus	MoodRing

and strategic drivers to ensure that they are current and unambiguous, restating/cross referring to them; (3) defining architecture principles; and (4) seeking commitment (among stakeholders) to the success of the architecture effort.

Based on (1)-(4), CEEADA approach can enable key stakeholders and the architect team to effectively collaborate when reviewing pre-existing business principles, goals, and strategic drivers. This will lead to a shared conceptualisation and understanding of significant enterprise aspects such as the enterprise mission, strategic plans, and external constraints among others. According to TOGAF, these are the key aspects for developing good architecture principles. A shared understanding will enable the determination of common criteria that will be used to evaluate architecture principles. Furthermore, a shared understanding will be a basis for defining architecture principles (i.e. identifying, elaborating, and validating elements of each architecture principle). Generated architecture principles can then be evaluated, such that adequate ones that echo business goals and strategic drivers are selected. Moreover, since gaining consensus on architecture principles is vital for the success of the architecture effort [32,24], CEEADA approach is useful because it focuses on gradually building consensus on various aspects when creating enterprise architecture.

5.2   Creating Architecture Vision

This TOGAF phase generally involves the following activities. (1) Seeking and gaining approval of the architecture project from corporate management, and

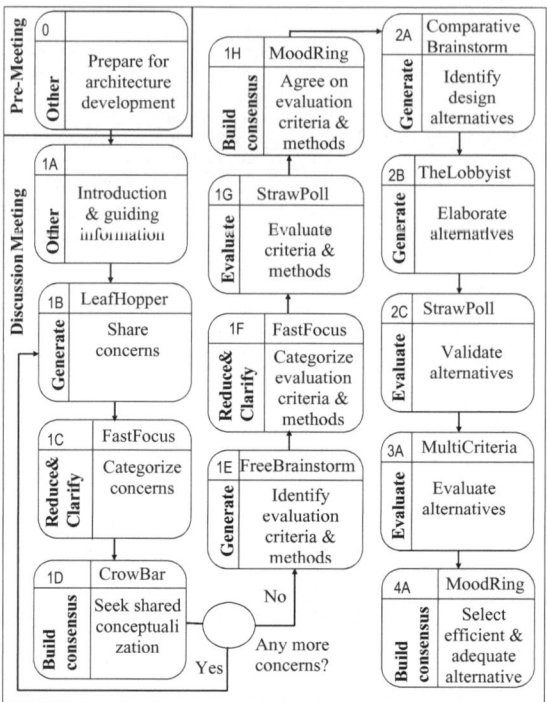

Fig. 4. Facilitation Process Model for CEEADA

commitment to its success from line management, (2) Identifying business goals and strategic drivers, or ensuring that their definitions (if pre-existing) are current and unambiguous, (3) Reviewing architecture and business principles, that will influence the development of the baseline architecture, ensuring that their definitions are current and unambiguous, (4) Defining the scope, and identifying and prioritizing the components of the baseline architecture. However, decisions regarding architecture scope should be made after practically evaluating the organisation's resources and competence, as well as the value that could be reaped if a given scope of the architecture work is chosen, (5) Defining enterprise-wide and project-specific constraints that the architecture must address, (6) Defining relevant stakeholders and their concerns, defining business requirements, and defining the high level description of the baseline and target environments that will address the requirements, within the defined scope and constraints, while conforming to business and architecture principles, and addressing stakeholders' concerns, and (7) Critically evaluating baseline environment, and documenting architecture vision in a statement of architecture work and seeking its approval.

Based on (1)-(7), CEEADA approach can enable key stakeholders and the architect team to effectively collaborate when reviewing and validating business goals, strategic drivers, business principles, and architecture principles. This will enable key stakeholders to acquire a shared conceptualisation and understanding of enterprise aspects significant for creating architecture vision. Moreover, evaluation criteria for possible solution alternatives in this phase can be determined. This is then followed by identifying, elaborating, and validating solution alternatives, i.e., architecture scope decisions, constraints, stakeholders' concerns, business requirements, components of the baseline and target (business, technology, data, and applications) architecture environments. Possible components of the baseline and target environments can then be evaluated, such that realistic and efficient ones are selected and consolidated into the statement of architecture work. According to TOGAF, consensus on the statement of architecture work determines the acceptability of the final architecture. Gaining consensus on the statement of architecture work is not a hassle if CEEADA approach is applied within this phase, because it will enable architects to gradually build consensus among stakeholders, when creating the architecture vision.

5.3 Business Scenarios: Business Requirements in the ADM

A business scenario *"is a description of a business problem in both business and architectural terms, which enables individual requirements to be viewed in relation to one another, in the context of the overall problem"* [3,32]. According to TOGAF, developing a business scenario involves Gathering, Analyzing, and Reviewing information on the following aspects. (1) The problem motivating the architecture effort, (2) the business and technical environments affected by the problem, (3) SMART (Specific, Measurable, Actionable, Realistic, Time-Sensitive) objectives to solve the problem, (4) human actors, and their places in the business model, (5) computer actors and computing elements, and their places in the technology model, and (6) responsibilities, success measures, and desired outcome for every actor.

Based on (1)-(6), CEEADA approach can enable key stakeholders and the architect team to effectively collaborate during the gathering phase. This will lead to an exhaustive gathering of business information on the aspects above, and enhance a shared conceptualisation and understanding of the such aspects. It will also enable architects to secure commitment from stakeholders. Furthermore, in the analysing phase where gathered information is filtered and translated into models, collaboration can be encouraged among key stakeholders. The idea of collaboration here is to enable identification and validation of possible alternatives regarding business requirements in order to address the problem. TOGAF literature highlights that in the reviewing phase, results of the analyzing phase are returned to stakeholders to seek a shared understanding of the problem scope and the possible depth of the technical impact. However, shared understanding can be steadily acquired if stakeholders are collaboratively involved in the early stages of developing business scenarios. Stakeholders should be involved in the filtering of gathered information on business requirements. This continuous involvement enables them to understand the reason(s) behind particular inferences in the business scenarios. The review phase could then be enriched by collaboratively evaluating the created business scenarios and selecting efficient ones.

Table 2. Summary of Insights from Walkthrough Sessions

#	CEEADA Aspects	Walkthrough 1	Walkthrough 2	Walkthrough 3
1	Prepare for architecture development sessions	– should not be a trivial activity – type of stakeholders involved affect the value of collaboration & evaluation of alternatives – The type of stakeholders to involve depends on scope of the organisation's problem – should include initial definition of organisation problem, & selection of stakeholders to involve in collaboration sessions – initial definition of problem scope initiates determining initial purpose of architecture effort, & preparation of stakeholders' concerns – all collaboration sessions should involve key decision makers of organisation units	– Architect team reveals calendar of events – Architect team briefs stakeholders on what they should expect from the architects, & what architects expect from stakeholders – Architects gain the trust of stakeholders – distribute agenda of a particular collaboration session prior to the session – all collaboration sessions should include key decision makers of organisation units	– determine the type of stakeholders to involve in every collaboration session
2	Introduction/ Briefing		– communicate purpose of the session & kind of information being sought for – get feedback on the agenda of a session	
3	Share concerns	– is successful if concerns were prepared by stakeholders prior to the session	– make explicit the type of concerns that stakeholders should share	
4	Categorize concerns			– clarify how to categorize concerns
5	Discuss concerns, seek shared conceptualisation & understanding of enterprise aspects	– Should seek for common understanding of organisation's problem scope, & initial purpose of the architecture effort, among other aspects		– Should also validate stakeholders' concerns against principles – valid concerns are vital for defining criteria & method for evaluating alternatives
6	Identify evaluation criteria & methods for alternatives	– is driven by the business goals to solve the organisation's problem		
7	Categorize criteria & methods	– instead validate criteria to be SMART		

6   Practice - Driven Insights into CEEADA

Constructed artifacts in design science are evaluated (using methods such as case study, action research, field study, and simulation among others) and the feedback obtained is used to refine the artifact further [11,12,13]. However, these artifacts must be tested in laboratory and experimental settings before field testing is undertaken [13]. In this research, before an experimental exploration of the performance of CEEADA models could be done, theoretical concepts in CEEADA had to first be validated by enterprise architects. Structured walkthrough sessions were used to expose these models to architects.

A walkthrough involves a step by step review and discussion, with practitioner(s), of activities that make up a process to reveal errors that are likely to hinder the effectiveness and efficiency of the process in realising its intended plan [17,19]. In addition to validating CEEADA models, walkthrough sessions were used to obtain industrial or practice-driven insights into our models. Three bi-lateral walkthrough sessions were conducted at Capgemini Netherlands, with three experienced enterprise architects. Architects who participated in the walkthroughs acknowledged the relevance of this approach in practice, and accordingly provided insights to improve the models.

Inputs to each session were figure 2 (CEEADA approach), and table 1 (task decomposition for CEEADA). Output from each session was feedback to improve

Table 3. Summary of Insights from Walkthrough Sessions - Continued

#	CEEADA Aspects	Walkthrough 1	Walkthrough 2	Walkthrough 3
8	Identify design or solution alternatives	– is driven by criteria balance – Should include stakeholders like business analysts, innovation department	– Architects may identify alternatives prior to session – Is hard to achieve in the case of principles. Architects compiles them – invite stakeholders to brainstorm on business requirements	– For the case of principles, architect compiles the list
9	Elaborate alternatives		– Indicate against each alternative, consequences (-ves & +ves) of choosing it. – In the case of business requirements, stakeholders should categorize them	– stakeholders help in the elaboration of principles
10	Validate alternatives	– effective & efficient if evaluation criteria are SMART – seeking for feasibility of alternatives	– seeking for feasibility of alternatives – stakeholders need to validate principles	– stakeholders need to validate principles
11	Evaluate alternatives	– Ranking, in the case of principles	– seeking quality of alternatives – In case of principles, stakeholders prioritize them – In case of architecture scope & constraints, negotiation dominates – In case of business requirements, stakeholders prioritize them	– for principles, stakeholders prioritize principles – Architect performs cross tabulation of principles against solution alternatives – architects consider relevance of opinion of @ stakeholder by assigning weights to them
12	Select efficient & adequate alternative	– may need to investigate candidate solution alternatives for more detail, before a final selection is done	– seek consensus on selected alternative(s)	– architecture board takes the decision (in the case of TOGAF ADM)

the models. The following three sections detail the analysis of feedback from the walkthroughs, and tables 2 and 3 summarise the output from all sessions.

6.1 Walkthrough Session One

The positive impact of collaboration between stakeholders and architects, and evaluation of enterprise architecture design alternatives depends on the type of stakeholders invited to the task. Stakeholders to participate in each collaboration session need to be carefully selected such that the right information is obtained and delays in making decisions, regarding deliverables of a session, are avoided. Moreover, the right stakeholders will be able to effectively and efficiently evaluate alternatives, and select appropriate and efficient design alternatives. It is therefore vital to indicate the type of stakeholders to be involved at each step of the proposed approach. For example key decision makers of the organisation units of interest should be involved in all steps of the proposed approach.

The type of stakeholders to be involved depends on the scope of the organisation's problem. The wider the scope, the higher you go up the rank of leaders; and the narrower the scope, the lower you go down the rank of leaders. Therefore, prior to step 1 in the proposed approach, a preliminary activity involving collaboration with senior management is vital. The idea for such an activity is to initially define the organisation's problem scope, and to select stakeholders who should participate in the subsequent collaboration efforts.

An initial definition of the organisation's problem scope, initiates the determination of the initial purpose of the architecture effort, as well as initial preparation of stakeholders' concerns. Thereafter seeking a common understanding among stakeholders, of both the organisation's problem scope and objective of the architecture effort, is indeed significant.

When defining common evaluation criteria and evaluation method for alternatives, architects should indeed collaborate with stakeholders. This is because business stakeholders have the expertise in evaluating and measuring quality of aspects in their business domain. Therefore, they should identify the possible evaluation methods, evaluate the identified methods, and then select a suitable one. The enterprise architect basically facilitates this activity and documents the aspects therein.

In practice, generation of design alternatives is driven by criteria balance. Therefore, it is vital to have explicit and valid evaluation criteria before generating design alternatives. Defining evaluation criteria for alternatives is driven by the organisation's problem scope and therefore business goals (e.g. swift cost reduction, swift volume growth, etcetera) to address the problem. Generation of alternatives is not the area of architects, so they should indeed collaborate with the stakeholders. Stakeholders that should be present may include business analysts, and process innovation department among others. In step 1 and 4 of the proposed approach, architects should facilitate the progress of the activities therein, while in steps 2 and 3, they should be actively involved as well as facilitate the associated activities.

Validation of alternatives for feasibility can be effective and efficient if the pre-defined evaluation criteria are SMART (Specific, Measurable, Actionable, Realistic, Time-sensitive). Additionally, depending on the phase of architecture development in which the approach is applied, investigating candidate solution alternatives for more detail before a final selection is done, could be vital. However, this may not apply in the case of architecture principles because the associated nature of evaluation is ranking of the principles.

6.2   Walkthrough Session Two

Stakeholders' concerns can be serious issues that could block the progress of the architecture work if not sufficiently addressed. Therefore, it is significant, to carefully address them when creating enterprise architecture. However, the term concerns as used in the proposed approach is ambiguous. In order to gather concerns exhaustively, there is need to specify the type of concerns that stakeholders should share during the collaboration sessions. Prior to the sessions, the architect team should draw a calender of events and organise an informal meeting with key stakeholders. In such a meeting, the team briefs stakeholders on what they should expect from them (the architects), and what the architects expect from the stakeholders, throughout the architecture creation process. This step is usually ignored by several architects yet it is crucial, because through such a gathering and clarification of events, it is very possible to gain the trust of the stakeholders.

The proposed approach can be useful during the high level specification of the architecture. However, during the collaboration sessions, it is essential to manage stakeholders' expectations, for example stakeholders know the agenda of a session before it begins. This enables them to make the necessary preparations for it. Moreover, before a collaboration session begin, its purpose, as well as the kind of information being sought for in that particular session, should be communicated. It is quite rewarding if architects identify some alternatives before the collaboration session of generating solution alternatives. This rules out the possibility of any associated difficulties amidst the session, and it helps to build confidence. Moreover, during the elaboration of identified alternatives, the consequences of choosing a particular alternative should be highlighted if possible. This fastens the validation and evaluation of the solution alternatives.

Depending on the phase of architecture development, architects often do the evaluation of alternatives and trade-off analysis without stakeholders. This affects the acceptability of the ultimate solution alternative. Yet seeking consensus on a chosen alternative is indeed significant. In every collaboration session, it is important to have key decision makers of the client organisation. For example if a decision is made in the absence of a CIO, this implies that in the next session when the CIO is present, if he does not agree with previously made decision; then activities must be repeated in order to make decisions in his support.

In practice it is difficult for architects to collaboratively generate architecture principles with the stakeholders. Architects commonly develop principles as follows. (1) They conduct interviews with senior management, (2) Findings from interviews are documented, and a list of architecture principles is compiled by the architects, (3) Principles are then presented to stakeholders for validation and prioritisation. Prioritising principles involves having stakeholders assign weights to them. It is easier having stakeholders prioritise principles, than generate them collaboratively with architects. Yet for the case of business requirements, stakeholders should be invited to brainstorm, categorise, and prioritise the requirements. Moreover, when defining architecture scope and constraints, negotiation is vital, rather than evaluation of design alternatives. The aspect of evaluating alternatives may arise during the negotiations.

6.3   Walkthrough Session Three

The categorisation of concerns in the process design should be clarified. For example since the approach is focusing on addressing collaboration related aspects in TOGAF ADM, categories of concerns should be specific to aspects in a particular TOGAF phase. This is because concerns are always related to objectives of a particular project. Stakeholders' concerns need to be validated before considering them in decision making. During the validation of a concern, the question of whether it matches principles should be answered. Valid concerns are useful for defining evaluation criteria, and choosing an evaluation method for alternatives. Since principles are always existent within the organisation but not written down, the architect collects information regarding the principles, and compiles it into a consistent set of about 10 to 15. The role of stakeholders then, is to validate and prioritise the principles.

In practice, when evaluating alternatives, the architect often performs a cross tabulation of principles against available alternatives. Each principle takes up

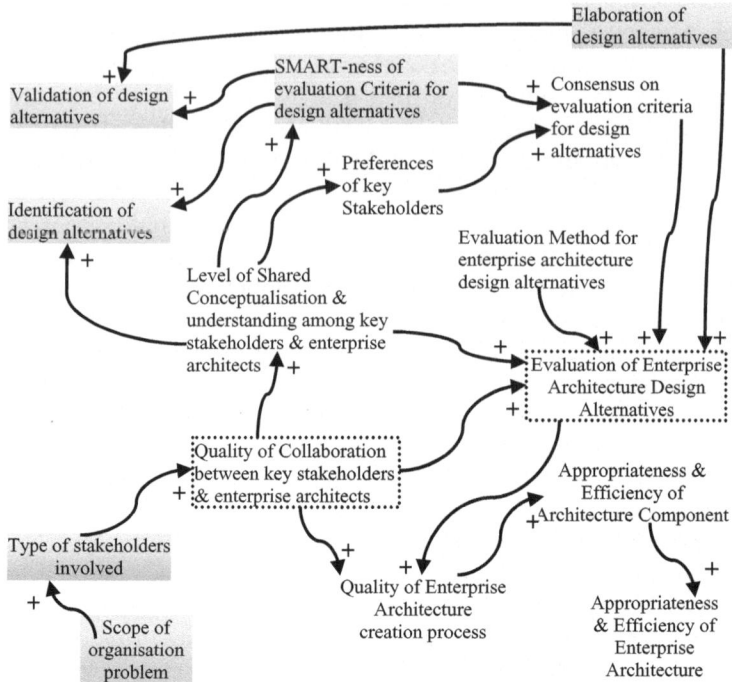

Fig. 5. Modified Cause-Effect Analysis in Creating Enterprise Architecture

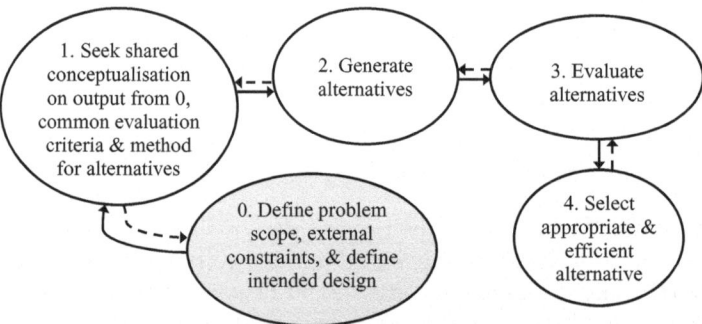

Fig. 6. Modified CEEADA Approach

a column in the table depending on its priority, while each alternative takes up a row. The performance of each alternative in fulfilling a given principle is assessed, and scores given. Moreover, during the prioritisation of principles and analysis of alternatives, architects must consider the relevance of opinion of each stakeholder. This is done by assigning weights to stakeholders. Documentation to justify judgements made on alternatives is also significant. To select alternatives, the architecture board (in the case of TOGAF) makes the final decision.

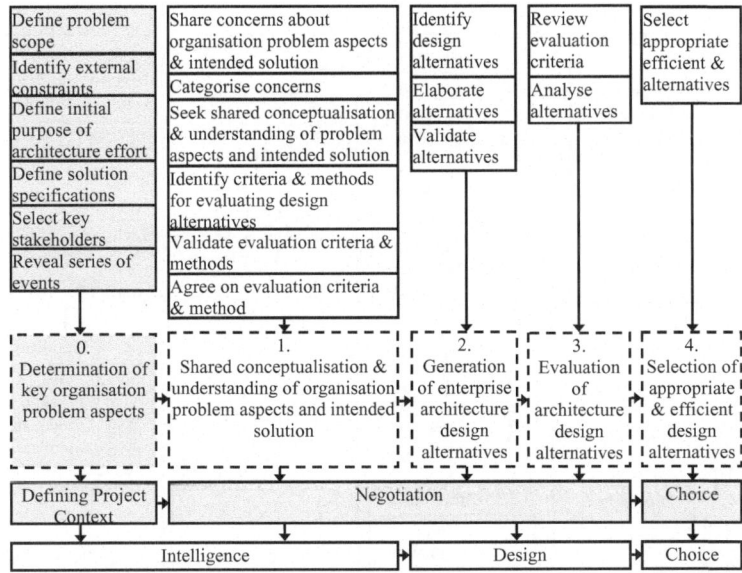

Fig. 7. Modified CEEADA Pattern Decomposition and Characterisation

Table 4. Modified Key Activities, Patterns of Collaboration, and ThinkLets

#	Activity Description	Deliverable	Stakeholders involved	Pattern of Collaboration, ThinkLet
0.1	Define initial organisation problem scope	Initial problem scope		-
0.2	Identify external constraints	Nonnegotiable constraints		
0.3	Define initial purpose of the architecture effort	purpose of the architecture effort	Senior management	
0.4	Select key stakeholders to participate in subsequent collaboration sessions	Key stakeholders to collaborate with architects		
0.5	Reveal calendar of events for architecture effort & expectations of architect team & key stakeholders	Calendar of events & expectations	All selected stakeholders	
	SESSION ONE – Seeking Shared Conceptualisation & Defining Common Evaluation Criteria			
1.1	Introduce purpose of session, kind of information required, organisation problem scope, & initial purpose of architecture effort	Guiding information		-
1.2	Stakeholder share concerns about initial purpose of the architecture effort & other aspects on organisation problem scope	Concerns	Decision makers of different organisation units	Generate, LeafHopper
1.3	Categorise concerns by type & organisation domains	Categories of concerns		Reduce & Clarify, FastFocus
1.4	Discuss concerns while seeking shared conceptualisation & understanding of problem aspects and initial purpose of architecture effort	Shared conceptualisation & understanding of problem aspects & architecture purpose		Build Consensus, CrowBar
1.5	Validate stakeholders' concerns	Valid concerns		Evaluate, StrawPoll
1.6	Agree on amendments to problem and solution aspects	Amendments to problem scope, and architecture purpose		Build Consensus, MoodRing
1.7	Identify criteria & methods for evaluating design alternatives	Evaluation criteria & methods		Generate, FreeBrainstorm
1.8	Validate criteria & methods	Valid criteria		Evaluate, StrawPoll
1.9	Agree on evaluation criteria & method for design alternatives	Common evaluation criteria & evaluation method		Build Consensus, MoodRing
	SESSION TWO – Generation of Enterprise Architecture Design Alternatives			
2.1	Introduction/Briefing	Guiding information	Business	-
2.2	Identify design alternatives	Design alternatives	analysts, process innovations unit,	Generate, ComparativeBrainstorm
2.3	Elaborate design alternatives	Elaborated design alternatives	IT architects, etc	Generate, TheLobbyist
2.4	Validate design alternatives	Validated design alternatives		Evaluate, StrawPoll
	SESSION THREE – Evaluation and Selection of Design Alternatives			
3.1	Introduction/Briefing	Guiding information	Decision makers	-
3.2	Evaluate valid design alternatives	Evaluated design alternatives	of organisation	Evaluate, MultiCriteria
4	Select appropriate & efficient design alternative	architecture design component	units	Build Consensus, MoodRing

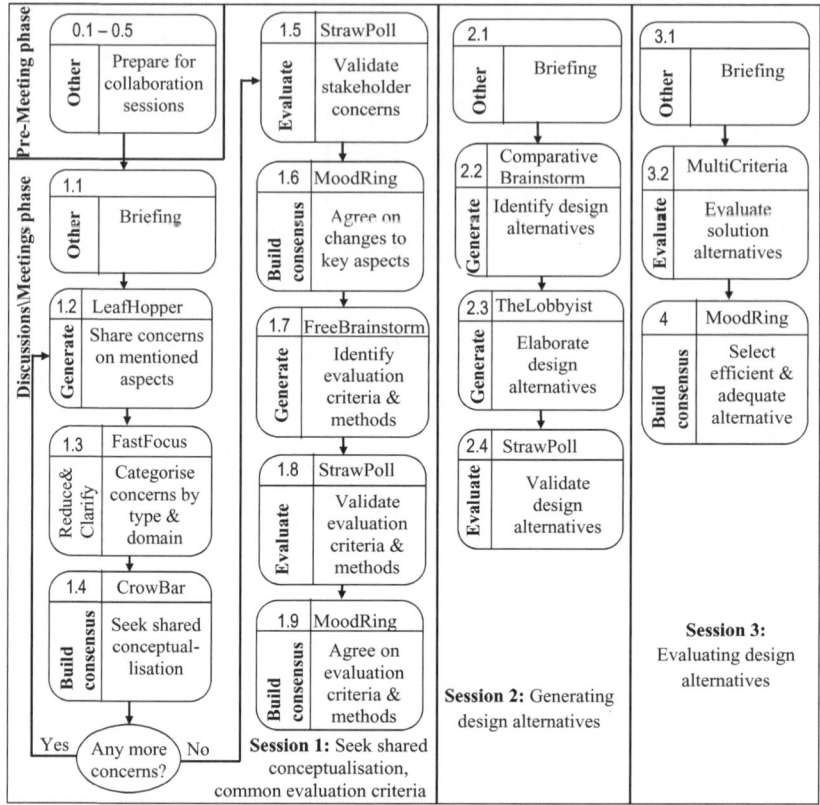

Fig. 8. Modified Facilitation Process Model for CEEADA

6.4   Revised CEEADA Models

Insights from the three walkthrough sessions were used to refine CEEADA models, i.e. the cause-effect analysis model (shown in figure 5), the pattern for CEEADA (shown in figure 6), and the collaboration process design for CEEADA (shown in table 4 and figure 8).

From the walkthroughs, other causal relations associated with quality improvement of the architecture creation process were obtained and amended (see shaded variables in figure 5). Explanations of these causal relations are given in section 6.

Figure 6 depicts an amendment of step 0 to the pattern for CEEADA. The relevance of step 0 is to enable enterprise architects with senior management to define the problem scope, identify external constraints from regulatory authorities, and define the purpose of the architecture effort. Key stakeholders to participate in the subsequent collaboration required in the architecture creation process, are also selected at step 0. These amendments arose from the walkthrough sessions (see section 6). Accordingly, in step 1 a shared conceptualisation and understanding of output from step 0 (i.e. problem scope, external constraints, purpose of the architecture effort, and solution specification) among other stakeholders is then appropriate. Iterativeness can also be identified within the pattern, in the sense that conflicts and errors that may arise in steps 2, 3, and 4, will be a result of ineffectiveness and inefficiency from steps 0 and 1.

Furthermore, figure 7 depicts modifications in the decoposition and characterisation of tasks in CEEADA. In the left part of figure 7, step 0 is decomposed into six tasks and characterised as part of Simon's intelligence phase. For the reason of making the underlying concepts of CEEADA more explicit and understandable, characterisation of CEEADA tasks has been further detailed (see bottom layers of figure 7). Step 0 is characterised as defining project context, steps 1-3 are characterised as tasks that involve negotiation, and step 4 is still characterised as choice.

As a result of modifications in figures 5, 6, and 7, the agenda plan for validating the collaboration process for CEEADA and its associated FPM, were modified as depicted in table 4 and figure 8 respectively.

7 Conclusions

In this chapter we presented theoretical underpinnings of CEEADA, an approach focusing on quality enhancement in creating enterprise architecture. The relevance of the approach in two phases of TOGAF ADM was discussed. In these phases, results of collaboration, negotiation and evaluation of design alternatives highly affect subsequent activities in the architecting effort. CEEADA models have been validated and enriched through structured walkthrough sessions with experienced enterprise architects. This resulted in modified models that represent both theoretical and practical insights into quality improvement of the architecture creation process. This chapter therefore contributes to efforts towards filling the gap (reported in [23]) of insufficient reflections on success factors for enterprise architecting.

Acknowledgements. We are extremely grateful to Richard Bredero, Karin Blum, Arnold van Overeem, and Claudia Steghuis, for their valuable contributions and practical insights into this research.

References

1. Armour, F.J., Kaisler, S.H., Liu, S.Y.: A big Picture Look at Enterprise Architectures. IT Professional, IEEE 1(1), 35–42 (1999a)
2. Armour, F.J., Kaisler, S.H., Liu, S.Y.: Building an Enterprise Architectures Step by Step. IT Professional, IEEE 1(4), 31–39 (1999b)
3. Blevins, T., Spencer, J.: The Open Group Architecture Forum: Manager's Guide to Business Scenarios. The Open Group (2002)
4. Briggs, R.O., de Vreede, G.J., Nunamaker Jr., F.: Collaboration Engineering with ThinkLets to Pursue Sustained Success with Group Support Systems. Journal of Management Information Systems 19, 31–64 (2003)

5. Briggs, R.O., Kolfschoten, G.L., Vreede, G.J., de Dean, D.L.: Defining Key Concepts for Collaboration Engineering. In: Proceedings of the Twelfth Americas Conference on Information Systems, Acapulco, Mexico (2006)
6. Chapurlat, V., Kamsu-Foguem, B., Prunet, F.: Enterprise Model Verification and Validation: an Approach. Annual Reviews in Control 27, 185–197 (2003)
7. Conklin, J.: Dialog Mapping: Building Shared Understanding of Wicked Problems. Wiley & Sons Limited, England (2006)
8. Figueira, J., Mousseau, V., Roy, B.: ELECTRE Methods. In: Figueira, J., Greco, S., Ehrgott, M. (eds.) Multiple Criteria Decision Analysis - State of the Art Survey. Springer, Heidelberg (2005)
9. Findeisen, W., Iastrebov, A., Lande, R., Lindsay, J., Pearson, M., Quade, E.S.: A Sample Glossary of Systems Analysis - Handbook of Applied Systems Analysis - IIASA. Web Dictionary of Cybernetics and Systems (accessed, February 16, 2009), http://pespmc1.vub.ac.be/ASC/ASCGloss.html
10. Gregor, S.: The Nature of Theory in Information Systems. MIS Quaterly 30(3), 611–642 (2006)
11. Hevner, A.R., March, S.T., Park, J., Ram, S.: Design Science in Information Systems Research. MIS Quarterly 28(1), 75–105 (2004)
12. Hevner, A.R.: A Three Cycle View of Design Science Research. Scandinavian Journal of Information Systems 19(2), 87–92 (2007)
13. Iivari, J.: A Paradigmatic Analysis of Information Systems as a Design Science. Scandinavian Journal of Information Systems 19(2), 39–64 (2007)
14. Janssen, M., Cresswell, A.: The Development of a Reference Architecture for Local Government. HICSS IEEE Press (2005)
15. Jonkers, H., Lankhorst, M.M., Doest, H.W., ter Arbab, F., Bosma, H., Wieringa, R.J.: Enterprise architecture: Management tool and blueprint for the organisation. Information Systems Frontiers 8(2), 63–66 (2006)
16. Kaisler, S.H., Armour, F., Valivullah, M.: Enterprise Architecting: Critical Problems. In: HICSS. IEEE Press, Los Alamitos (2005)
17. Kolfschoten, G.L., de Vreede, G.J.: The Collaboration Engineering Approach for Designing Collaboration Processes. In: Haake, J.M., Ochoa, S.F., Cechich, A. (eds.) CRIWG 2007. LNCS, vol. 4715, pp. 95–110. Springer, Heidelberg (2007)
18. Kolfschoten, G.L., Briggs, R.O., Appelman, J.H., de Vreede, G.J.: ThinkLets as Building Blocks for Collaboration Processes: A Further Conceptualization. In: de Vreede, G.-J., Guerrero, L.A., Mar n Raventos, G. (eds.) CRIWG 2004. LNCS, vol. 3198, pp. 137–152. Springer, Heidelberg (2004)
19. Kolfschoten, G.L.: Theoretical Foundations for Collaboration Engineering. Delft University of Technology, The Netherlands (2007)
20. Lankhorst, M., van Drunen, H.: Enterprise Architecture Development and Modelling, http://www.via-nova-architectura.org
21. Muller, G.: How to relate design decisions to stakeholder satisfaction: Bridging the broad stakeholder universe and the detailed technology world (2007), http://www.via-nova-architetura.org
22. Nakakawa, A.: Collaboration Engineering Approach to Enterprise Architecture Design Evaluation and Selection. In: Proceedings of 15th CAiSE-DC (Doctoral Consortium) held in conjunction with CAiSE 2008. CEUR-WS, Montpellier, France, vol. 343, pp. 85–94 (2008)
23. Op't Land, M., Proper, H.A(E.), Waage, M., Cloo, J., Steghuis, C.: Enterprise Architecture - Creating Value by Informed Governance. Springer, Berlin (2008)

24. Richardson, G.L., Jackson, B.M., Dickson, G.W.: A Principles-Based Enterprise Architecture: Lessons from Texaco and Star enterprises. MIS Quarterly 14(4), 385–403 (1990)
25. Rouwette, E.A.J.A., Vennix, J.A.M.: System Dynamcis and Organisational Interventions. Systems Research and Behavioral Science 23, 451–466 (2006)
26. Schekkerman, J.: The Economic Benefits of Enterprise Architecture, How to quantify and Manage the economic Value of Enterprise Architecture. Trafford Publishing, Canada (2005)
27. Schekkerman, J.: How to survive in the jungle of Enterprise Architecture Frameworks, Creating or Choosing an Enterprise Architecture Framework. Trafford Publishing, Canada (2004)
28. Schelp, J., Stutz, M.: A Balanced Scorecard Approach to Measure the Value of Enterprise Architecture (2007), http://www.via-nova-architectura.org
29. Simon, H.A.: The New Science of Management Decision. Harper and Row, New York (1960)
30. Simon, H.A.: The Sciences of Artificial, 3rd edn. The MIT Press, Cambridge (1996)
31. Slot, R.: What is the ROI of Architecture? Reporting on the added value of architecture. Landelijk Architectuur Congrees (2004), http://www.lac2004.nl/docs/fvbg2hdsb83/Sponsors/Capgemini.pdf
32. TOGAF - The Open Group Architecture Framework Version 8.1.1 Enterprise Edition (2007), http://www.togaf.org
33. van den Bent, B.: A Quality Instrument for Enterprise Architecture Development Process. Master's Thesis Business Informatics, Utrecht University (2006)
34. Van der Raadt, B., Schouten, S., Van Vliet, H.: Stakeholder Perspective of Enterprise Architecture. In: Morrison, R., Balasubramaniam, D., Falkner, K. (eds.) ECSA 2008. LNCS, vol. 5292, pp. 19–34. Springer, Heidelberg (2008)
35. Veltman-van, R.E.: Determining the Quality of Enterprise Architecture Products. Master's Thesis Business Informatics, Utrecht University (2006)
36. Vreede, G.J., de Briggs, R.O.: Collaboration Engineering: Designing Repeatable Processes for High-Value Collaborative Tasks. In: HICSS. IEEE Press, Waikoloa (2005)
37. Extensible Architecture Framework version 1.1 (format edition), report of the NAF working group (2003), http://www.naf.nl/content/bestanden/xaf-1.1-fe.pdf

Architecture-Driven Requirements Engineering

Wilco Engelsman[1], Henk Jonkers[1], Henry M. Franken[1], and Maria-Eugenia Iacob[2]

[1] BiZZdesign, P.O. Box 321, 7500 AH Enschede, the Netherlands
[2] University of Twente, School of Management and Governance, P.O. Box 217, 7500 AE Enschede, the Netherlands

Abstract. This paper presents an architecture driven requirements engineering method. We will demonstrate how to integrate requirements engineering in architecture design and we will demonstrate how to use enterprise architectures during solution realization projects. We will demonstrate how architecture can be used during problem investigation, solution specification and solution validation through an example application.

Keywords: Requirements Engineering, Goal Oriented Requirements Engineering, Enterprise Architecture.

1 Introduction

In the last few years the enterprise architecture (EA) paradigm emerged to better adapt organizations to changing customer needs [25] [9]. EA is believed to increase the understanding of scope for solution realization projects. Solution realization projects in this context are the projects executed to reach the to-be architecture, for example introducing a new business service.

Requirements Engineering (RE) as a scientific discipline has matured over the last decade. The process itself is rather well understood and has led to numerous techniques and models (e.g. GBRAM [1], I* [24] KAOS [19] and more traditional techniques like interviews, workshops [3] or viewpoint oriented RE [10]). However, we believe that enterprise architectures can have a tremendous impact on the requirements engineering process. Not only do requirements lead to architecture, there is also much progress to be made in constraining the requirements process with the relevant scope, context and structure. Lastly eliciting and specifying requirements from architectural models can give the requirements engineers a head start before traditional techniques like workshops and scenario based elicitation come into play. We believe that there is much progress to be made by clearly defining the relationships between RE and EA. To explore our ideas on the proposed integration of architecture into requirements engineering we performed an exploratory case study at a large Dutch insurance company to extract this information. In this paper we will demonstrate how requirements engineering leads to architecture and architecture leads to requirements. This distinction can be made because architecture is either a design artifact or a frame of reference. We will position requirements engineering in both these views and propose a way of integrating these views.

E. Proper, F. Harmsen, and J.L.G. Dietz (Eds.): PRET 2009, LNBIP 28, pp. 134–154, 2009.

The structure of this paper is as follows, in section 2 we will demonstrate our view on requirements engineering. In sections 3 and 4 we provide the theoretical boundaries of architectures and RE and describe the conclusions from our study. In section 5 we provide a framework for architecture-driven requirements engineering and in section 6 we provide an example application of our method. In chapter 7 we provide an outlook for further research.

2 Requirements Engineering

Requirements Engineering (RE) is involved with investigating and describing the environment in which the envisioned system is supposed to create desired effects and designing and documenting behavior of the envisioned system [3]. In other words, RE is about getting from problems to the possible solutions. In general, there might be more than one valid or needed solution for a particular problem. Each solution itself can be another problem for someone else (see figure 2). This was recognized by Jackson as a progression of problems [7].

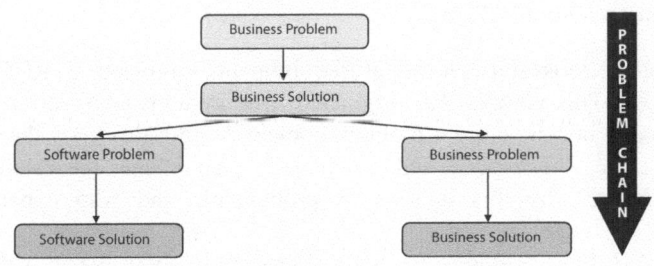

Fig. 1. A progression of problems

We define RE as getting from problems to the possible solutions. We do not limit ourselves to technology based solutions. Therefore we define a solution as a system that provides desired services. A system can be a new information service to customers, new business processes, new work procedures, supporting software systems and application services that support business processes.

Because RE is about bridging the gap between a problem and the possible solutions, two different views on RE have emerged [2] [21] [22]: *problem-oriented* and *solution-oriented* RE. Problem-oriented RE is about problem investigation: to investigate and determine what the actual problem is. Problem-oriented RE involves finding and documenting the problematic phenomena before thinking of how to solve that particular problem. A key concept in this is: understanding the goals and the stakeholders who experience these goals. Solution-oriented RE is about designing and describing system behavior and showing which alternative best solves the problem. If we try to relate these two views, we can argue that problem-oriented RE and solution-oriented RE come together in *architecture* [21]. If we investigate a certain problem, for example, we determined that certain stakeholders experience that the service delivery to customers is insufficient to realize certain business goals. Then a solution to this problem could be new business processes, a new delivered service to the customer, alignment of existing business processes to this new service and a new supporting software system. Together these different solutions form the architecture of an overall solution (see figure 2). We will elaborate this statement in chapter 4 and 5.

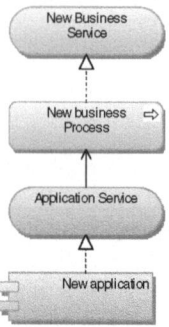

Fig. 2. Overall solution architecture

2.1 Problem-Oriented RE

In this section we look more closely at Requirements Engineering (RE) as a problem solving activity. This view on RE originated from systems engineering and is about investigating and documenting a problem domain. Within this view the requirements engineer describes the experienced problematic phenomena, the relations between these phenomena, why this is seen as problematic and who experiences these problems.

Wieringa [2] [21] provides us with information about what a Requirements Specification (RS) should contain when RE is seen as problem analysis; a RS in this view describes the desired business objectives and what work should be done to reach these business objectives. A similar distinction can be found in Tropos [4]. Tropos uses an early and late requirements phase, where the early requirements phase describes the system objectives and the late requirements phase describes the functional and non-functional requirements.

A very popular RE technique within problem-oriented RE is Goal Oriented RE (GORE). GORE [1] has received a large amount of research efforts over the past years and its popularity has increased ever since. Goals are regarded as high-level objectives of the business, organization or system. They capture the reasons why a system is needed and guide decisions at various levels within the enterprise. For a general description about GORE in practice see the work of Van Lamsweerde [18]. Relevant work in the field of GORE has been done by the authors of GBRAM [1], KAOS [19] and I* [24]. The main reason to adopt a GORE based approach is the inadequacy of traditional system approaches (e.g. structured analysis or object oriented analysis) to capture the actual motives for the system under development. Traditional approaches treat requirements as consisting only of data and processes and do not capture the rationale for the systems, making it difficult to understand high-level concerns in the problem domain.

2.2 Solution-Oriented RE

This view on requirements engineering (RE) is the traditional software engineering view on requirements engineering. When using the view of solution specification a requirements specification consists of [21]:

- A specification of the context in which the system will operate.
- A list of desired system functions of the system.
- A definition of the semantics of these functions.
- A list of quality attributes of those functions.
- A Demonstration which alternative best solves the problem.

Traditional techniques in this view are structured analysis [15] and object-oriented analysis [8]. Object-oriented analysis applies techniques for object-modeling to analyze the functional requirements for the system under development. Structured analysis focuses on the data that flows through the system under development.

3 Enterprise Architecture and Requirements Engineering

Enterprise Architecture (EA) is the complete, consistent and coherent set of methods, rules, models and tools which will guide the (re)design, migration, implementation and governance of business processes, organizational structures, information systems and the technical infrastructure of an organization according to a vision [6]. In this chapter we will discuss relevant Enterprise Architecture frameworks and their views on Requirements Engineering (RE).

3.1 TOGAF

In popular methods for enterprise architecture, such as The Open Group Architecture Framework (TOGAF, see figure 3)[17], (business) goals and requirements are central drivers for the architecture development process. In TOGAF's Architecture Development Method (ADM, see [17]), requirements management is a central process that applies to all phases of the ADM cycle. The ability to deal with changes in the requirements is crucial to the ADM process, since architecture by its very nature deals with uncertainty and change, bridging the divide between the aspirations of the stakeholders and what can be delivered as a practical solution.

TOGAF provides a limited set of guidelines for the elicitation, documentation and management of requirements, primarily by referring to external sources. TOGAF's content meta-model, part of the content framework, defines a number of concepts related to requirements and business motivation; however, this part has been worked out in little detail compared to other parts of the content meta-model, and the relation with other domains is weak. Also, the content framework does not propose a notation for the concepts. We do recognize the fact that requirements engineering drives architecture design. But TOGAF lacks the distinction between architecture as a design artifact and architecture as a frame of reference. In the former architecture is the result from RE, the latter uses architecture as a frame of reference to guide RE.

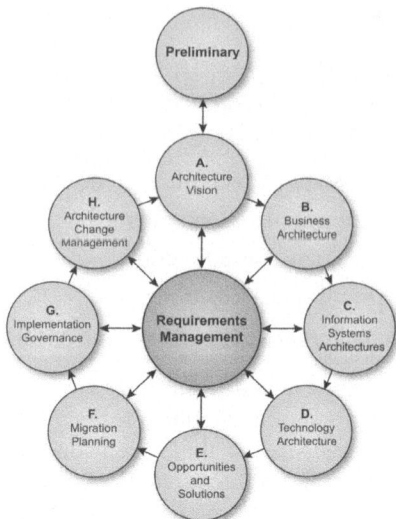

Fig. 3. TOGAF ADM

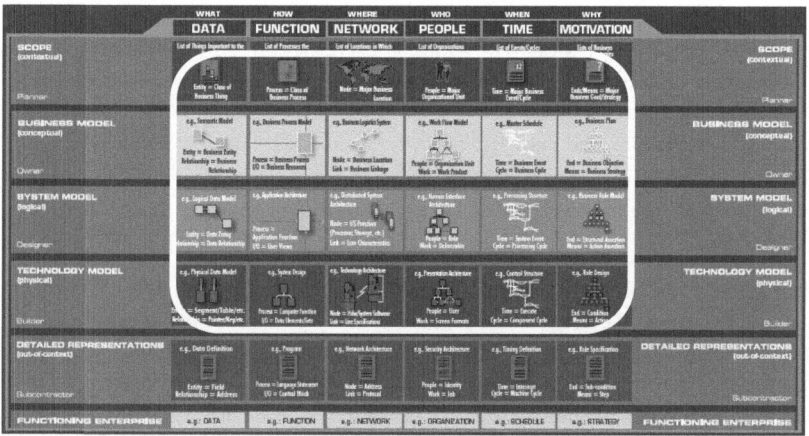

Fig. 4. Zachman framework

In our view problem-oriented requirements engineering drives architectures design. This architecture design is then a solution to the experienced problems. This was already established by Michael Jackson and his problem frame approach [7] through a progression of problems.

3.2 Zachman Framework

The ancestor of Enterprise Architectures is the Zachman framework [26]. The framework as it applies to enterprises is simply a logical structure for classifying and organizing the descriptive representations of an enterprise that are significant to the management of the enterprise as well as to the development of the Enterprise's systems. It was derived from analogous structures that are found in the older disciplines of Architecture/Construction and Engineering/ Manufacturing that classify and organize the design deliverables created during the process of designing or producing complex physical products.

Figure 4 presents and overview of the "Framework for Enterprise Architecture", usually known as the Zachman Framework. An important aspect in this framework is the motivation column. The motivation column explains why the architecture is needed. Looking at the motivation column we can already distinguish a link to requirements engineering. Explaining the motivation for either the architecture or the elements of the different architectural layers can be realized through problem-oriented RE.

3.3 Project Start Architecture

A technique to align architectures with solution realization is Project Start Architecture (PSA) PSA [12] describes the relevant parts of the reference architecture at the start of a project. The PSA is a steering instrument that ensures the relevancy of the architecture in concrete projects. Architecture should not be an academic exercise by architects, but of concrete value in organizational change.

The PSA is a translation of general principles and guidelines relevant for the change projects. Relevant parts of the (reference) architecture are selected and written down in the PSA. The PSA is then handed down to the relevant project for solution realization. The solution realization process uses the PSA as an input document and validation document. The solution should use the boundaries set by the PSA. The PSA provides the context of solution realization; it does not describe the solution itself. The idea of PSA is very useful as it transfers scope and frame of reference to solution realization projects. It clearly defines the boundaries of the problem under investigation and solution designers can use this scope to specify detailed solution behavior.

3.4 I* for Enterprise Architecture Design

Eric Yu [14] [23] proposes to use I* as a problem investigation technique for architecture design and business modeling. This way the motivation for architectural elements is linked to their implementation. I* [24] is a technique that focuses on modelling and reasoning support for early phase requirements engineering. It tries to capture the understanding of the organizational context and rationales that lead up to systems requirements. It consists of two main modelling components. The Strategic Dependency (SD) model is used to describe the dependency relationships among various actors in an organizational context. The Strategic Rationale (SR) model is used to describe stakeholder interests and concerns, and how they might be addressed by various configurations of systems and environments [24].

I* can be used for both early and late phases of RE. During the early requirements phase $i*$ is used to model the environment of the system to be, it facilitates the analysis of the domain by allowing the modeller to diagrammatically represent the stakeholders of the system, their objectives and their relationships. During the late phases i* models are used to propose the new system and the new processes and evaluate them on how well they meet the functional and non-functional needs of the users.

4   Alignment Architecture with Requirements Engineering

When analyzing the relevant literature from chapter 3 we can conclude the following on architecture-driven requirements engineering. Requirements play an important role in architecture design. We argue that the requirements for architecture design are problem oriented and the architecture provides the general design to these goals. This way architecture is seen as a design artifact, the solution for identified problems. Secondly we need to transfer the motivation and the architecture design into the solution realization projects. Here architecture is the frame of reference. Transferring the relevant parts of the architecture is the domain of Project Start Architecture (the solution itself or the solution blueprints are not in the domain of PSA). Although the documentation and theoretical integration into RE is weak (e.g. see [12]). To further investigate this we performed an exploratory case study, unfortunately we are unable to provide exact details of this case study due to confidentiality reasons. We can only provide some context in which the case study took place. To investigate our claims we elicited requirements from an off the shelf reference architecture and compared these to the results from traditional requirements elicitation techniques. This was done through eliciting requirements from the reference architecture based on the project goals. These requirements were compared to the results from the actual project. This way we were able to compare requirements elicited from architectural models with requirements from traditional techniques. We were able to show that we can use architecture to assist the elicitation and analysis of requirements, improve requirements specification and help validate the requirements.

4.1   Requirements Elicitation and Analysis

During requirements elicitation we were able to elicit a large number of requirements by inspecting the reference architecture. Furthermore, the traditional approach and the architecture-driven approach led similar requirements. The architecture-driven requirements were more general in nature. The reference architecture provided a scope for the problems and described possible solutions for these problems. We will elaborate this with an example. The company were the case study took place faced a problem about the integration of a new product in their current insurance portfolio. This new product required the use of an insurance broker. At the time of integrating this new product into the company they only sold insurances directly to their customers. This triggered a new problem since they had no idea how to implement and realize an insurance broker distribution channel and how to provide IT support for an insurance broker administration. Their architecture described this for them. For example, during a workshop a requirement emerged that for an insurance broker his name, address, bank account number and chamber of commerce data had to be recorded. When investigating the reference architecture we saw reference models describing recording insurance broker information and abstract examples of this information.

A second observation was that the architecture-driven approach facilitates the re-use of similar solutions as source for requirements elicitation. For example, when a new insurance product (or service) is introduced similar products could be used to elicit change requirements. Lastly the architecture provided the relevant scope for the new systems through analyzing relationships between architectural elements.

One point of consideration is that the company used an off the shelf reference architecture. This architecture already described a desired to-be state, without a to-be analysis. For example, it provided solutions for problems that were not experienced yet. It therefore also provided ready to use solutions for problems. Secondly this way the organization had a very mature architecture to begin with, so it was quite easy to transfer the relevant models into solution realization projects.

4.2 Requirements Specification

During our case study we introduced a requirement specification template based on the architecture framework used in the organization. This template consists of business, application and technology layers (based on the meta-model used in their organization). This template was used to specify the requirements elicited from the architecture. The main argument to develop a template for a requirements specification around this meta-model is that solving an organizational problem is much more than just investigation and specifying the IT need. In section 2 we explained that our view includes a progression of problems. Using this template we were able to show which business problems were solved on the business layer and their relationship to the application level. Furthermore, we were able to show how specifying the requirements for a business service impacts and serves as input for specifying requirements for the supporting information systems. Thus, one could emphasize the underlying dependency relationships between the requirements positioned in the different layers of the above-mentioned template.

4.3 Requirements Validation

During requirements validation, the role of the enterprise architect is similar to that of any other stakeholder during the validation phase. In this setup the architect is regarded as stakeholder in the validation activities and may judge whether the specified requirements comply with the architecture goals, guidelines, principles, policies and constraints and with the architecture. Secondly, since the architecture described a desired to-be state it provided a validation mechanism in the form that a requirements specification should comply with.

5   Architecture-Driven Requirements Engineering

We have established that Requirements Engineering (RE) both happens to design architectures and realize the architecture. To design the architecture RE investigates the problematic phenomena, describes the business objectives and a way of working to realize these objectives. To realize the architecture we need to transfer the relevant requirements for the architecture and the architecture design into the solution realization projects. This way we heavily restrain the freedom of the solution designers to match the already established architecture.

5.1   Framework for Requirements Engineering

We argued that Requirements Engineering (RE) is about getting from problems to the possible solutions. Therefore we use a logical framework for problem solving [20] (see figure 5) as a RE framework.

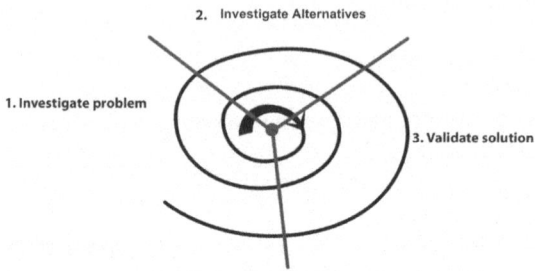

Fig. 5. Framework for requirements engineering

5.1.1   Problem Investigation

During problem investigation we take the problem-oriented view on requirements engineering. We find the stakeholders, record the relevant business objectives and specify how to reach these business objectives. This phase uses the concepts introduced in GORE (see KAOS [19] and $i*$ [14] [23]). This phase in our method leads to a goal tree that serves as input for the traditional requirements techniques. Important concepts during this phase are the stakeholders, their concerns, assessments of these concerns, goals (both hard and soft goals) and requirements. A precise definition and report on the design of the requirements language is out of the scope of this paper. But we will provide an exact syntax, to elaborate the example (see table 1).

5.1.2   Investigate Alternatives

In this step we start to look for possible solutions that are available to solve our problem. Solution specification is an important activity during this phase. Solution designers [5] propose system properties during this phase to reach the goals identified earlier. Solution alternatives range from proposing new (business) systems to actual alternative solution properties.

5.1.3   Solution Validation

In the solution validation phase the different solution alternatives are compared and analyzed. The main goal is to determine which solution best implements the business requirements [13]. Another important goal in this activity is to identify new problems. For example, when we have identified the need for a new service that we wish to provide to our customer and specified its desired behavior we are imposed with another problem. How are we going to realize this service internally? Other needed solutions might be adapted business processes, new information systems and a changed infrastructure. This leads to another cycle of the RE method.

Table 1. Elements from the requirements language

Abstract element	Concrete notation
Stakeholder	Stakeholder
Concern	Concern
Assessment	Assessment
Hard goal	Hard goal
Soft goal	Soft goal
Requirement	Requirement
Use case	Use case

5.2 Framework for Architecture-Driven Requirements Engineering

In this framework (see figure 6) architecture is either a design artifact which requires requirements engineering or a frame of reference which guides requirements engineering.

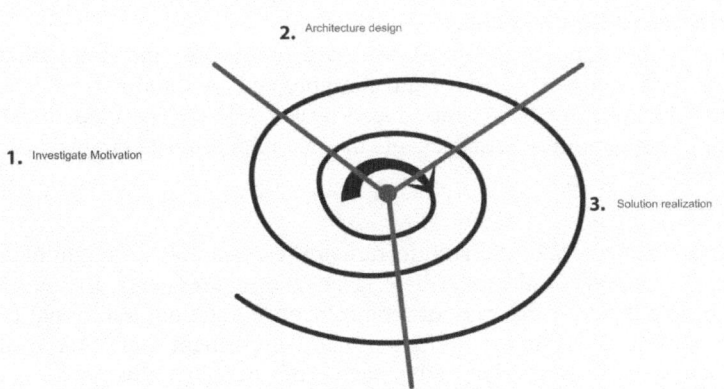

Fig. 6. Life cycle for architecture-driven RE

The framework from figure 5 applies to the framework from figure 6 as well. For example, we find the steps problem investigation, investigate alternatives and validate solution in the individual steps of this framework as well.

5.2.1 Investigate Motivation

Investigate problem

Before architecture design we investigate the motivation for the architecture. In requirements engineering terms, we investigate the business objectives and the way of working to reach these objectives. In the early stages, before architecture design we propose to use problem-oriented requirements engineering. More concretely, we have adopted a goal oriented approach. Goals are an excellent mechanism to explain the motivation for a solution [18]. If we compare this to existing GORE approaches, it resembles the early requirements phase found in i* [24].

Investigate alternatives

During investigate alternatives we propose the solutions for the particular problems depicted in the motivation plane. We also start specifying the solutions on a high level. For example initial use-case specification models. It is not required to provide detailed use-case specifications during this phase. When the solution designers start working on the solution they can take these use-case specifications as a starting point.

Validate solution

During solution validation the proposed solutions are evaluated and new problems are investigated. These relationships define the progression of problems defined by Jackson [7]. Solution validation during this phase focuses more matching the proposed solution to the goals and identifying new problems on an architectural level. For example in this setup, the stakeholder concerned with validation activities may judge whether the specified requirements comply with the enterprise architecture goals, guidelines, principles, policies and constraints [11].

5.2.2 Solution Realization

During solution realization we transfer the solutions from design plane and their motivation to the realization projects.

After the architecture is designed we need to transfer the motivation and the architectural models to the solution realization projects. A solution for this is found in Project Start Architecture (PSA) introduced by DYA [12]. The models defined here should lead to a blueprint of requirements that the requirements engineers can use for their solution specification.

Problem investigation

We now know what parts of the architecture are relevant and we might have solution blueprints. The architectural model here steers the requirements elicitation process. When we have exhausted this way of requirements elicitation, traditional techniques, like workshops and scenario elicitation can supplement our first draft of the requirements specification. The advantages of working this way is that the requirements elicitation activities get a head start and are constrained by the relevant parts of the organization, depicted in an architectural model. Architecture helps the requirements engineer with elaborating the relevant scope of the problem under investigation. For example, when we know that the change goal for our project is to develop a new business service that allows customers to administer and maintain their insurance portfolio over the internet. The architecture can then provide the relevant models for insurance products, insurance selling processes, etc.

Investigate alternatives
During this phase we take a much more traditional approach. We investigate alternatives constrained to the solution we have to realize.
We use solution specification techniques for detailed solution specification. In terms of an IT system think of techniques from Object Oriented Analysis or Structured Analysis. These techniques are found in solution oriented requirements engineering. For business solutions, specification techniques from the business domain could be used. For example service blueprinting for a business service.

Solution validation
During solution validation we compare the different possible solutions to the system objectives. Validation is about to show which solution is expected to reduce the gap between the experienced problems and the desires.

6 Example

In this section we will provide an example case for our requirements engineering method. We will demonstrate how to use architectural models during problem investigation, solution specification and solution validation.

> PRO-FIT is an average sized financial service provider, specialized in different insurance packages, such as life insurances, pensions, investments, travel insurances, damage insurances and mortgages.
>
> In the last years PRO-FIT went through a structural change process, the result of which is that all business processes are consistent up the department level However, the financial branch is one of the most dynamic and the senior management of PRO-FIT is now aware of new developments and threats, which require PRO-FIT to think of new ways to deal with these new challenges
>
> During the identification of new developments and threats the senior management of PRO-FIT became aware of the new service-oriented way of thinking. A market analysis identified a number of opportunities; one of them is a differentiation strategy for their insurance services using modern technology.
>
> During the past few months the customer support at PRO-FIT identified a number of problems as well. Customers are complaining about the lack of insight in their insurance portfolios, competitors offer new internet based solutions where customers can request all kinds of information about their insurance portfolios.

During the remainder of this example we will use the requirements language depicted in table 1.

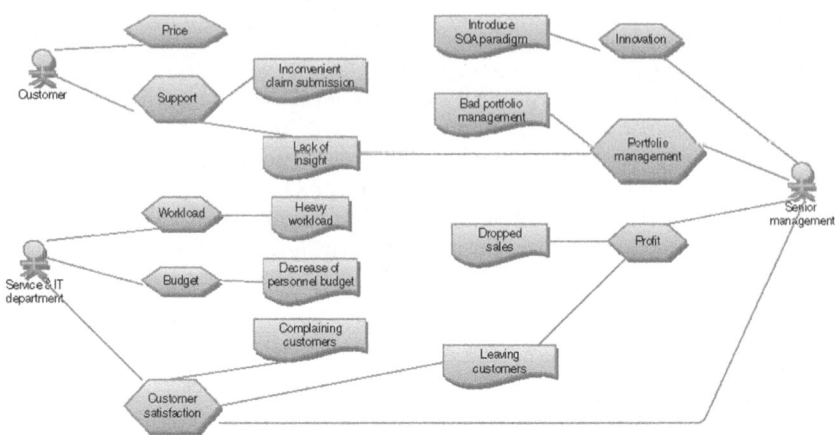

Fig. 7. The stakeholders, concerns and assessments

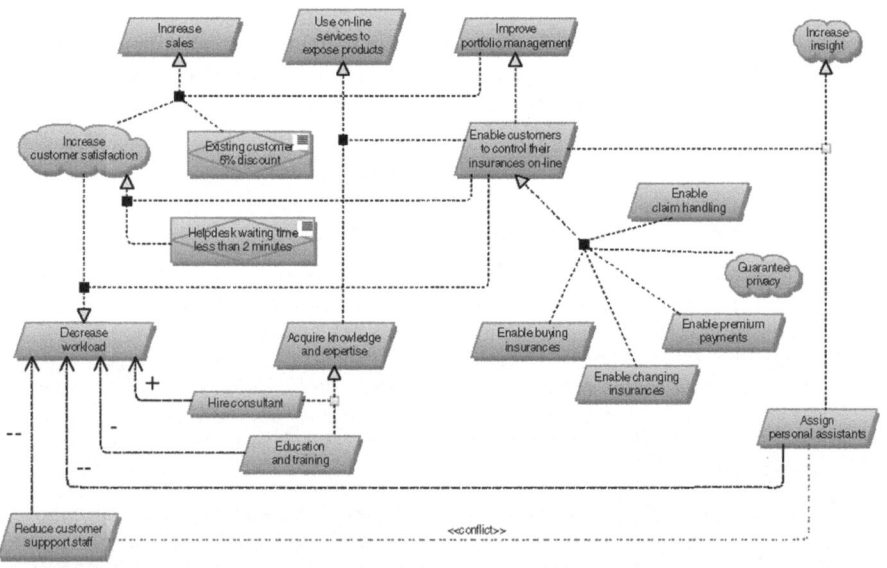

Fig. 8. Results from problem oriented requirements engineering

6.1 Investigate Motivation

During this step we will investigate the motivation of PRO-FIT. We will explore the stakeholders, their concerns and assessments. These concerns and assessments lead to goals. In this step architecture is a design artifact.

6.1.1 Investigate Problem

Because of space limitations we will restrict this example to three relevant stakeholders, with a limited number of concerns and assessments. We assume that we have a customer who is concerned with price and support. We also have a stakeholder (or stakeholders) senior management. Senior management is concerned with innovation, portfolio management and profit. Thirdly, we have identified the customer service department as a stakeholder. They are concerned with workload, budget and customer satisfaction. See fig. 6 for an overview of the concerns and assessments.

As we can see the identified concerns from the respective stakeholders can lead to assessments. These assessments are ways to address these concerns, for example the concern profit leads to an assessment of dropping sales. This is a threat to the organization and therefore needs to be addressed. This will lead to the high level goal "increase sales" (see figure 8). Through goal refinement we reach the goals that we want to introduce a new portfolio management service that allows the customer to buy insurances online, mutate his/her data online, pay their premiums and submit their claims.

6.1.2 Investigate Alternatives

During investigate solution alternatives we investigate the possible solutions which will realize our goals from section 5.1. In our case we will introduce a new portfolio management service. We use use-case specification to specify high level behavior. The use case portfolio management describes the high level behavior and can be refined into refined use-cases (see figure 9).

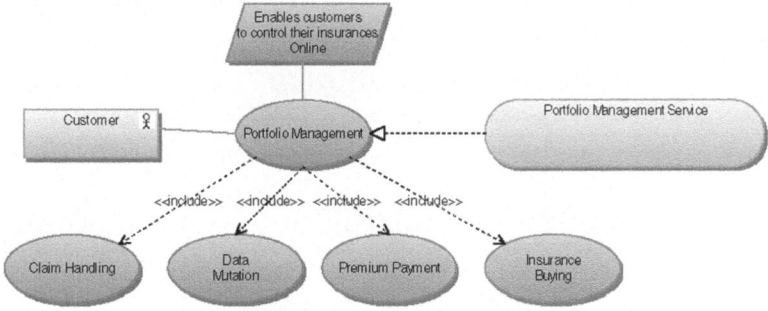

Fig. 9. New Portfolio Management Service

6.1.3 Validate Solution

During solution validation we both check the current specified solution and try to identify new problems. In this case it is determining the IT support. The solution defined in this chapter is then a problem for the IT specialist. In the next cycle of problem investigation and solution specification PRO-FIT assumes the role of a service consumer. During solution validation the architecture can be used to identify new problems based on the proposed solution. For example, during this example we introduced a new business service. This business service might introduce new business processes and it will need IT support. One way of finding new problems is to perform an impact analysis on the architecture [11].

6.2 Solution Realization

We argued that before solution realization starts, the architecture should be inspected for relevant information. This coincides with "architecture as a frame of reference". In a best case scenario the architecture already describes a to-be state; this to-be state already provides a number of requirements and seriously limits the solution alternatives. When there is no to-be state, the architecture still provides relevant models, scope, context and structure to the RE activities. The relevant parts of the reference architecture comprise of guidelines, principles and the relevant models found. In this section we will provide a selection of the relevant models from the PRO-FIT architecture. We will realize a more business oriented solution, but this way of thinking also applies for IT based systems.

Because we know that we want to sell insurances we can select the product architecture for product information. We can also select the processes "claim handling" and "new insurance request". Using the business information model we can already select the relevant information requirements. The reference architecture also describes PRO-FITS insurance portfolio, namely car insurances, life insurances, travel insurances and general liability insurances.

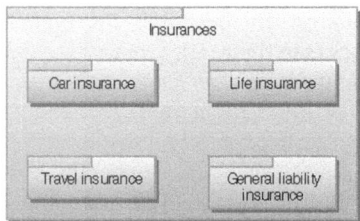

Fig. 10. The product architecture of PRO-FIT

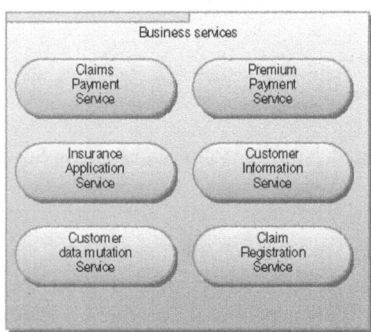

Fig. 11. The business service architecture of PRO-FIT

In figure 6 we illustrate the business services PRO-FIT delivers to its customers or internal departments.

Processes

In the BIP phase guidelines where given that the new service should support selling insurances, changing insurances, claim handling and premium payments. For illustration purposes we selected the relevant process models for closing contracts and claim handling. Closing contracts is the internal procedure for handling insurances requests.

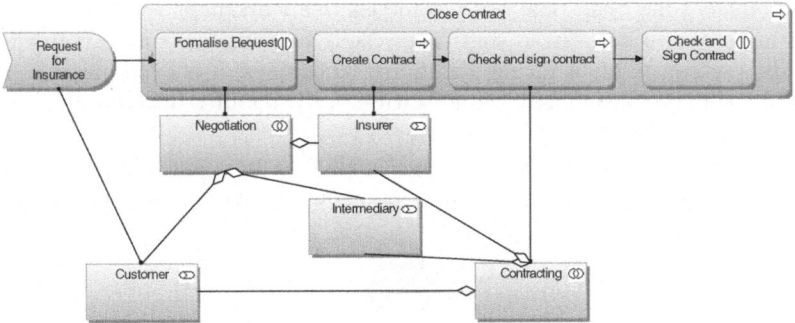

Fig. 12. Close Contract business process including the relevant entities and roles

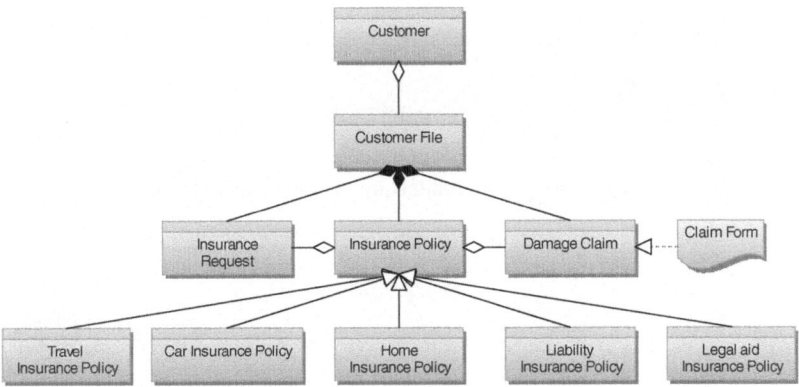

Fig. 13. Business information in PRO-FIT

Business Information

The architects also need to provide the relevant business information parts of the architecture. They already identified the relevant products and processes. In we depict PRO-FITS current business information model, as recorded in the reference architecture.

Another important aspect is to transfer the motivation for the architecture. The solution designers have to use the scope identified for the architecture. Secondly the use-cases specified at the architecture restrain the scope for the solution designers.

6.2.1 Problem Investigation

During problem investigation the architecture can provide the requirements engineer with the relevant models to determine the scope. After the business information planning phase we know the main goals and that PRO-FIT wants to realize a new portfolio service. Through asking ourselves how and why questions we can refine the goal tree from figure 3. First sources of information are the architectural principles depicted in figure 4. The client satisfaction goal should be used for every solution realization project. The business function model also provides the relevant refinement goals (or relevant process models). An architecture driven way of working does not mean it replaces the traditional soft techniques like workshops and interviews. It is a supporting phase to get a head start. The results from architecture driven elicitation should be used as an input for the traditional techniques. It is even possible to refine the goal "support insurance selling" with "sell liability insurances", "sell car insurances" using the product architecture.

During the solution realization we can also elicit requirements that realize "provide security". Supporting "user identification" is a goal that refines "provide security".

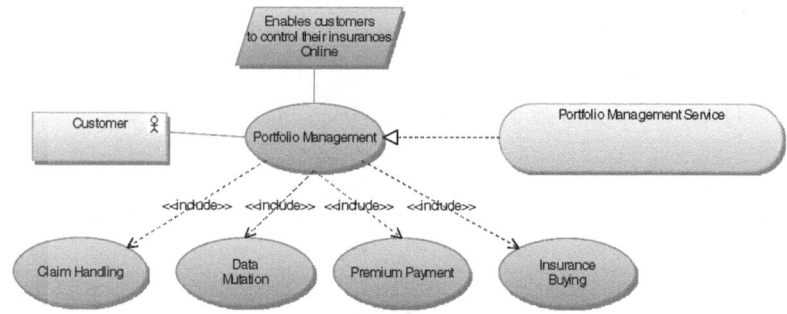

Fig. 14. Reuse of architecture solution specification

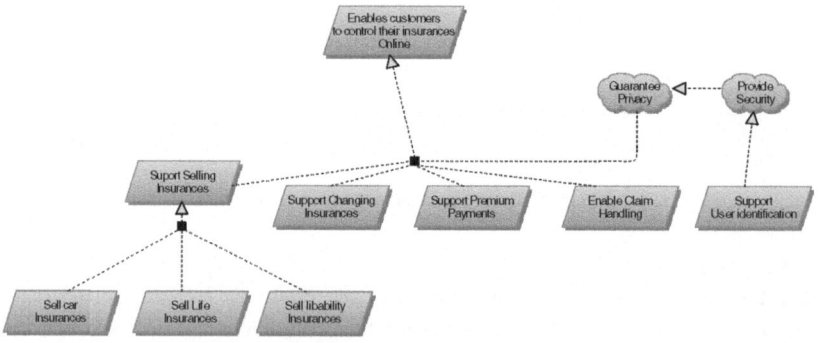

Fig. 15. Extension of the motivation

As mentioned before the exact details of these requirements are made clear using the traditional techniques. As we saw with our case study, the architecture provides less detailed requirements. Situational details should be elicited the old fashioned way. Another example is the relevant business information. Inspecting the architectural model from figure 13 provides the layout of the business information. Adding details to the objects is still required.

6.2.2 Investigate Solution Alternatives

In this step the person responsible for specifying the solution investigates the possible alternatives. For example in figure 16 he identifies two requirements that realize the goal "support user identification". He has two possibilities here, using either Digi-ID or some form of biometric identification.

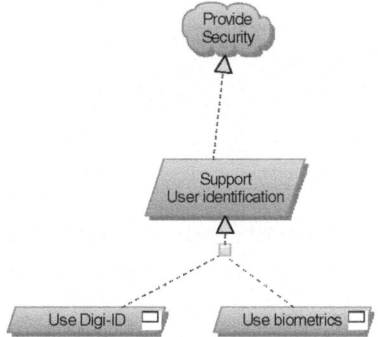

Fig. 16. Determining solution properties

In figure 17 we see the requirements "use i-deal" or "use credit cards" are possible alternatives to realize "support premium" payments.

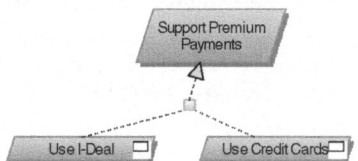

Fig. 17. Solution alternatives for support premium payments

A second step during this phase is specifying solution behavior. In figure 18 we demonstrate how the earlier use-cases from the previous phase are refined into more concrete solution behavior.

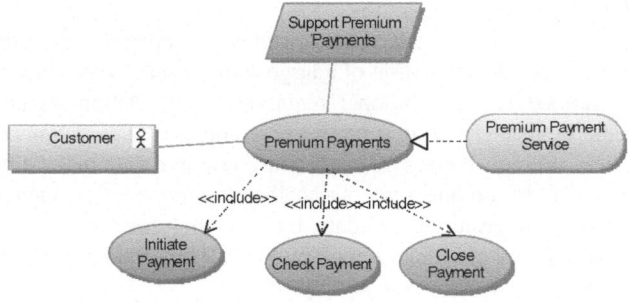

Fig. 18. Solution specification for premium payment Service

Solution specification does not end here. In figure 13 we provided the business information model for PRO-FIT. These initial data requirements can then be supplemented using the traditional techniques like workshops, interviews and scenario based elicitation.

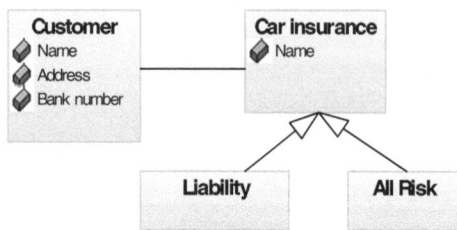

Fig. 19. Adding details to the object models from the architecture

6.3 Solution Validation

During solution validation we both check the current specified solution and try to identify new problems. In this case it is determining the IT support. The solution defined in this chapter is then a problem for the IT specialist. In the next cycle of problem investigation and solution specification PRO-FIT assumes the role of a service consumer. During solution validation the architecture can be used to identify new problems based on the proposed solution. For example, during this example we introduced a new business service. This business service might introduce new business processes and it will need IT support. One way of finding new problems is to perform an impact analysis on the architecture [11].

7 Concluding Remarks and Future Work

In this paper we have described the influence of the Enterprise Architecture (EA) paradigm on the way in which Requirements Engineering (RE) is performed. An extensive survey and classification of existing literature has shown that the link between these two areas is still weak. For a large part, the results described in this paper are based on the observations made during a practical case study carried out within a large Dutch insurance company.

In the first part of the paper, we have shown that a company's enterprise architecture can be a useful source for the elicitation of a large starting set of requirements. These may subsequently be refined using traditional requirements elicitation techniques, such as scenarios, workshops, interviews or surveys. This approach has a number of potential advantages: (1) time savings, among others because requirements may be reused between different projects; (2) the architecture places the requirements in their organizational context, which makes it easier to validate them with business stakeholders; (3) the architecture provides a way to structure requirements, which makes it easier to check for quality aspects such as consistency and completeness.

In the second part of the paper, we have made the combined approach to EA and RE operational by proposing a method for architecture-driven requirements engineering. This method includes a process (way of working) and concepts for modeling requirements and their relationship to other concepts in the enterprise architecture. The method has been illustrated with a practical example.

As future work, we intend to fully validate our method in the pilot project we are currently carrying out. Although we have shown that it is possible to elicit requirements from enterprise architectures, we still do not know exactly how much improvement architecture-driven requirements engineering can actually offer. For example, how much of solution specification can we realize based on results from an architecture-driven elicitation process? How much faster is an architecture driven approach?

Secondly we need to extend the framework described here with analysis possibilities. For example, the stakeholder concerns are similar to the viewpoints in viewpoint oriented RE [16]. Identifying standard viewpoints or methods for viewpoint identification is a logical next step.

Another interesting topic for future research is the relationship between service-oriented computing and requirements engineering. The ideas from service orientation may further facilitate the reuse of requirements and solutions, thus speeding up the requirements engineering phase. However, service-oriented solutions may also lead to change in the requirements engineering process. In particular, we envisage that separate (complementary) requirements engineering processes are needed for the service provider and the service user.

Acknowledgements

This work wouldn't have been possible without the input from Dick Quartel, in the form of ARMOR, the requirements modeling language used in this paper.

References

[1] Anton, A.I.: Goal-based requirements analysis. In: Proceedings of the Second International Conference on Requirements Engineering, pp. 136–144 (1996)

[2] Aurum, A., Wohlin, C.: Engineering And Managing Software Requirements. Springer, Heidelberg (2005)

[3] Bray, I.K.: An Introduction to Requirements Engineering. Addison-Wesley, Reading (2002)

[4] Bresciani, P., Perini, A., Giorgini, P., Giunchiglia, F., Mylopoulos, J.: Tropos: An agent-oriented software development methodology. Autonomous Agents and Multi-Agent Systems 8(3), 203–236 (2004)

[5] Cross, N.: Strategic knowledge exercised by outstanding designers. Strategic knowledge and concept formation III, pp. 17–30 (2001)

[6] Iacob, M.E., Franken, H., Van den Berg, H.: Enterprise Architecture Handbook. Bizzdesign academy publishers (2007)

[7] Jackson, M.: Software requirements & specifications: a lexicon of practice, principles and prejudices (1995)

 [8] Jacobson, I.: The use-case construct in object-oriented software engineering. Scenario-based Design: Envisioning Work and Technology in System Development, 309–338 (1995)
 [9] Jonkers, H., van Burren, R., Arbab, F., de Boer, F., Bonsangue, M., Bosma, H., ter Doest, H., Groenewegen, L., Scholten, J.G., Hoppenbrouwers, S., et al.: Towards a language for coherent enterprise architecture descriptions. In: Proceedings of Seventh IEEE International Enterprise Distributed Object Computing Conference, 2003, pp. 28–37 (2003)
[10] Kotonya, G., Sommerville, I.: Requirements engineering with viewpoints. Software Engineering Journal 11(1), 5–18 (1996)
[11] Lankhorst, M.: Enterprise Architecture at Work: Modelling, Communication and Analysis. Springer, Heidelberg (2005)
[12] Luipers, J.: White paper project start architectuur (in dutch). Sogeti internal report
[13] Mulholland, A., Macaulay, A.L.: Architecture and the integrated architecture framework. Whitepaper, Capgemini,
     `http://www.capgemini.com/services/soa/ent_architecture/iaf/`
[14] Samavi, R., Yu, E., Topaloglou, T.: Strategic reasoning about business models: a conceptual modeling approach. Information Systems and E-Business Management, pp. 1–28
[15] Schoman, K., Ross, D.T.: Structured Analysis for requirements definition. IEEE Trans. on Software Engineering 3(1) (1977)
[16] Sommerville, I., Sawyer, P., Viller, S.: Viewpoints for requirements elicitation: a practical approach. In: Proc. Third IEEE International Conference on Requirements Engineering (ICRE 1998) (1998)
[17] The Standish The Open Group. Togaf™ version 8.1.1 enterprise edition (2006),
     `https://www.opengroup.org/architecture/togaf8-doc/arch/`
[18] van Lamsweerde, A.: Goal-oriented requirements engineering: a roundtrip form research to practice. In: Proceedings of 12th IEEE International Requirements Engineering Conference, 2004, pp. 4–7 (2004)
[19] van Lamsweerde, A., Letier, E.: From object orientation to goal orientation: A paradigm shift for requirements engineering. In: Wirsing, M., Knapp, A., Balsamo, S. (eds.) RISSEF 2002. LNCS, vol. 2941, pp. 325–340. Springer, Heidelberg (2004)
[20] Wieringa, R.: Requirements Engineering: Frameworks for Understanding. Wiley, Chichester (1996)
[21] Wieringa, R.: Requirements engineering: Problem analysis and solution specification (extended abstract). In: Koch, N., Fraternali, P., Wirsing, M. (eds.) ICWE 2004. LNCS, vol. 3140, pp. 13–16. Springer, Heidelberg (2004)
[22] Wieringa, R., Heerkens, H.: Requirements engineering as problem analysis: Methodology and guidelines. Technical report, University of Twente (2003)
[23] Yu, E., Strohmaier, M., Deng, X.: Exploring Intentional Modeling and Analysis for Enterprise Architecture (2006)
[24] Yu, E.S.K., Mylopoulos, J.: Understanding why in software process modelling, analysis, and design, pp. 159–168 (1994)
[25] Zachman, J.A.: Enterprise Architecture: The Issue of the Century. Database Programming And Design 10, 44–53 (1997)
[26] Zachman, J.A.: A Framework for Information Systems Architecture. IBM Systems Journal 38(2/3), 454–470 (1999)

Informed Governance of
Enterprise Transformations

Frank Harmsen[1,2], H.A. Erik Proper[1,3], and Nicolette Kok[1]

[1] Capgemini, Utrecht, The Netherlands
[2] Maastricht University, Maastricht, The Netherlands
[3] Radboud University Nijmegen Nijmegen, The Netherlands

Abstract. Modern day enterprises are confronted with a variety of challenges, forcing them to continuously transform themselves to better meet these challenges. The diversity of the challenges and the resulting desires to transform (parts of) the enterprise, make it desirable to align all required and desired transformations in such a way that they complement each other rather than nullifying, or even undermining, each other's effects. Therefore, mature governance of these enterprise transformations is absolutely crucial. We will argue that this requires a *transformation authority*, being the organisational function which is responsible for the governance of enterprise transformations. In this chapter, our driving interest is the implementation of mature *transformation authorities*.

Judging whether a portfolio of enterprise transformations is well aligned requires insight into the desired overall result, as well as the planned and *achieved* effects of the individual transformations. This is what we refer to as *informed governance*. In this chapter we will position the discipline of *enterprise architecture* (referring to the architecture of the enterprise, and not just enterprise-wide IT architecture) as the core means to achieve informed governance. We will argue that mature governance of enterprises transformations presupposes the use of enterprise architecture to direct the portfolio of transformations.

Our discussions we will be based on theories from management science, as well as experiences from our own industrial practices. We will also discuss two cases of enterprises involved in the implementation of *transformation authorities* and use these to further refine our theoretical model.

1 Introduction

As a result of developments such as globalisation, the fusion of business and IT, the introduction of new technologies, novel business models, et cetera, enterprises are confronted with an increasing variety of options to deal with an ever faster changing environment. This results in a need for enterprises to be able to innovate, and to adapt themselves quickly to these changes in the environment, as well as a desire to proactively exploit these developments in an attempt to create new business opportunities. As a result, modern day enterprises are confronted with several challenges driving them to continuously transform themselves to put them in a position where they are better equipped to meet these challenges.

The diversity of the challenges and the resulting desires to transform (parts of) the enterprise, make it necessary to align all required and desired transformations in such

E. Proper, F. Harmsen, and J.L.G. Dietz (Eds.): PRET 2009, LNBIP 28, pp. 155–180, 2009.

a way that they complement each other rather than nullifying, or even undermining, each other's effects. This also puts a major challenge on an enterprise's management to make the right decisions at the right time and ensure that these decisions are translated into the right actions. At the same time, since enterprise transformations are executed in terms of projects, one needs to ensure that these projects comply to the decisions made. In practice, this proves to be a difficult task indeed. Even more, different stakeholders and/or problem-owners will have a different perception of the necessary changes and their priority. Unless properly governed, chaos will result. Mature governance of enterprise transformations is therefore absolutely crucial, requiring a dedicated *transformation authority* as the organisational function responsible for the governance of these enterprise transformations. In this chapter, our driving interest is the implementation of mature *transformation authorities*.

Judging whether a portfolio of enterprise transformations is well aligned, requires insight into the desired overall result as well as the planned and *achieved* effects of the individual transformations. This is what we refer to as *informed governance*. In line with [1], we position enterprise architecture[1] as the core means to achieve *informed governance*. This will be elaborated upon in Section 4, where we will argue that mature governance of enterprise transformations presupposes the use of enterprise architecture to direct the portfolio of transformations.

In our discussions we will take both a theoretical perspective, basing ourselves on theories from management science [2,3] and cybernetics [4,5] as well as a practical perspective based on experiences from industrial practice. More specifically, we also discuss two cases of enterprises aiming to implement *transformation authorities*. We will use these cases to further refine our theoretical model.

The remainder of this chapter is structured as follows. In Section 2, we start with a brief exploration of drivers for organisations to transform themselves. We will see how these drivers may actually pull the enterprise into different directions. Stakeholders with a stake in the outcome of the transformation, may want to direct the enterprise's transformation in different directions based on the discussed drivers. This begs for the implementation of an *transformation authority*. Before we can properly discuss the concept of a *transformation authority*, we need to define more specifically what we mean by *enterprise transformations* and their *governance*. Therefore, Section 3 provides a theoretical exploration of these concepts. In Section 4 we then position enterprise architecture as a necessary means for informed governance of portfolios of enterprise transformations. Using these definitions, Section 5 then identifies the requirements to be put on a *transformation authority*, as well as the processes involved in its maturation.

With our, initial, theoretical framework in place, we then proceed in Section 6 and 7 by discussing two (anonymised) cases drawn from industrial practice involving the implementation/maturation of a *transformation authority* in a pre-existing large organisation. In Section 8 we provide an analysis of the two cases in relation to the initial theoretical framework. An important conclusion from this analysis will be that the implementation of a transformation authority requires a broad maturity framework taking several important aspects into consideration that may lead to blockages during its

---

[1] We understand enterprise architecture as the architecture of the enterprise, and not as a synonym for enterprise-wide IT architecture.

implementation, and maybe even lead to erosion of already achieved results. This leads to the introduction of a refined theoretical framework in terms of a *transformation maturity framework* (TMF) in Section 9.

2   Drivers for Enterprise Transformation

This section is concerned with a brief exploration of the drivers which may trigger organisations to transform themselves. These drivers are likely to pull the enterprise into different directions. Stakeholders with a clear stake in the outcome of the transformation, may also want to direct the enterprise's transformation in different directions based on the drivers discussed. Without an effective governance mechanism making a clear univocal choice for future direction, and ensuring that the transformation stays on course. The discussed below is based on a more elaborate discussion provided in [1].

2.1   Keep Up or Perish

Enterprises face many changes, such as mergers, acquisitions, innovations, novel technologies, new business models, reduced protectionism, de-monopolisation of markets, deregulation of international trade, privatisation of state owned companies, increased global competition, etcetera. These changes are fuelled even more by the advances of eCommerce, Networked Business, Virtual Enterprises, Mashup Corporations, the availability of resourcing on a global scale, et cetera [6,7,8,9]. These factors all contribute towards an increasingly dynamic environment in which enterprises want to thrive.

2.2   Shifting Powers in the Value Chain

Clients of enterprises have become more demanding. A shift of power in the value chain is occurring. Clients have grown more powerful and demand customised, integrated and full life-cycle products and services. For example, rather than asking for a "printer", they require a guaranteed "printing service". Even more, customers have a tendency to ask for integrated service offerings. Rather than treating booking of a ight, a hotel, and a sight-seeing trip as separate services provided via separate outlets, customers opt for one-stop shopping. A shift from basic products to full services.

The creation and delivery of such complex products and services requires additional competencies which may not be readily available within a single (pre-existing) enterprise. In this pursuit they increasingly engage in complex product-offerings involving other parties, leading to cross selling and co-branding. To ensure the quality of such products and services, a high level of integration and orchestration between the processes involved in delivering them is required.

2.3   Comply or Bust

In the networked economy, governance of enterprises becomes increasingly complex. One sees a shift in governance from individual departments within an organisation, to the entire organisation, and lately to the organisation's value web. Management does not only have to worry about the reputation of their own organisation, but also about the other organisations in their value web.

How daunting the latter might be can be illustrated by real life examples, such as a large shoe manufacturer who outsourced the production of shoes to another company, to only discover at a later stage that the latter made use of child labour. Although the latter company was not part of the shoe maker's own organisation, their reputation was still damaged, threatening their survival on the market-place.

Corporate governance is not only an issue to an organisation on its own, but also a major concern to society as a whole. As a result of undesired and uncontrollable effects of the increased socio-economical complexity and interdependency of organisations, services, products and nancial instruments. Recent examples of such side-effects are the well-known Enron scandal, as well as the sub-prime mortgage crises. To control and/or prevent such effects, new legislation has been put in place to better regulate enterprise practices. An example being the Sarbanes-Oxley Act [10] forcing enterprises to increase the quality of their governance and appropriateness of audits.

2.4   Achieving Competitive Advantage

Enterprises try to achieve and maintain a competitive advantage. In order to do so, they need to choose an optimal strategic position. Porter [11] distinguishes four basic units of competitive advantage: product development, purchasing, operation, and distribution of products or services. Performing these four activities better than rivals do is called operational excel lence. Enterprises can, however, also opt for other ways of distinguishing themselves from their rivals. In [12], Treacy and Wiersema argue that enterprises should try and focus on one of the three disciplines of added value: product leadership, operational excel lence and customer intimacy.

In the recent past, enterprises needed to excel only in one of the above areas to be successful, and meet industry standards on the other areas [12]. Due to the network economy and globalisation, there is a growing need to excel in a minimum of two areas (or at least in one and signicantly increasing in the other areas).

2.5   Making Technology the Business Differentiator

The evolution of information technology brings an abundance of new opportunities to enterprises. Technology becomes part of almost everything and most processes have become IT reliant, if not fully automated. The technological evolutions confront enterprises with the question of which technologies are relevant to the enterprise? Which technology should be replaced and which technology could be of use for developing new products (or services) of to enter new markets?

2.6   Excel or Outsource

Increasingly enterprises outsource business processes. Outsourcing of business processes requires organisations to precisely understand and describe what needs to be outsourced, as well as the implementation of measures to ensure the quality of the outsourced processes [13,14,15,16].

In deciding on what to outsource and how to safeguard its quality, management needs insight into the extent to which processes can be outsourced, the risks that may need to be managed when doing so, as well as the interdependencies within the outsourced processes and between the outsourced processes and the retained organisation.

Conversely, organisations with a strong tradition in a certain business process may decide to become industry leader for such processes. For example, processing of payments, management of IT infrastructure and logistics.

3   Governed Enterprise Transformation

In this section we provide a theoretical perspective on enterprise transformations and in particular the governance thereof. In doing so, we will base ourselves on theories from management science [2,3] and cybernetics [4,5], as well as our own experiences from industrial practice.

We regard an enterprise as an (open and active) system comprising a collective of actors, processes and technology which jointly engage in some *purposeful* activity. Being a system in the *general systems theory* sense of the word [4,5], an enterprise can be divided into component systems (such as business units) as well as aspect systems (such as IT, business processes, et cetera).

An enterprise may evolve over the course of time. This evolution may be the result of a gradual change of the behaviour of individual elements in the enterprise, or it may be the result of a deliberate and conscious action. We define an enterprise transformation as the latter type of change, in other words, a deliberate and conscious action aiming to make changes to an enterprise. This is illustrated in Figure 1. At the execution level we find the operational enterprise concerned with "normal" operations (a *first order* system), while at the transformation level we find the enterprise transformation (a *second order* system).

Enterprise transformations may be triggered by several events. Management of an enterprise needs to make conscious decisions about the initiation and direction of enterprise transformations, balancing the desired benefits of the transformation in relation to its costs. Note: a special type of enterprise transformation would be the 'undoing' or 'prevention' of unwanted gradual change of the enterprise. More importantly, this also requires a conscious decision about the desiredness of the direction the 'natural evolution' takes and the need to counter or stimulate this. Furthermore, as argued before, when a series of transformations is executed, these transformations need to be aligned. The enterprise transformation and the operational enterprise may be sub-systems of the

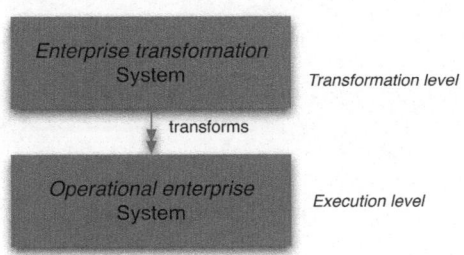

Fig. 1. Enterprise transformation

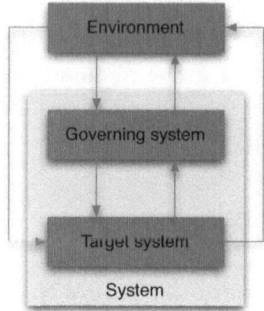

Fig. 2. The basic governance paradigm

same enterprise, but they could equally well be part of different enterprises. In the first case, the enterprise is able to execute its own transformations, while in the second case an external party (e.g. a consultancy firm) is used to execute the transformation.

According to [17], governance is *"the activity of [] controlling a company or an organisation"*. In management science this is embodied by the so-called governance paradigm [2,3]. Figure 2, which is based on [2], depicts the basic governance paradigm. The governance paradigm involves three important assumptions:

1. there is some system, and not as a synonym to application system as is the case in software development. In the context of enterprise architecture, we are specifically interested in active systems [18], the target system, which interacts with its environment;
2. this target system needs to be governed;
3. there is another system, the governing system which does the actual governing.

The essence of the governance paradigm is that during the execution of a process (the *target system*) there is some kind of interaction with the environment (input and

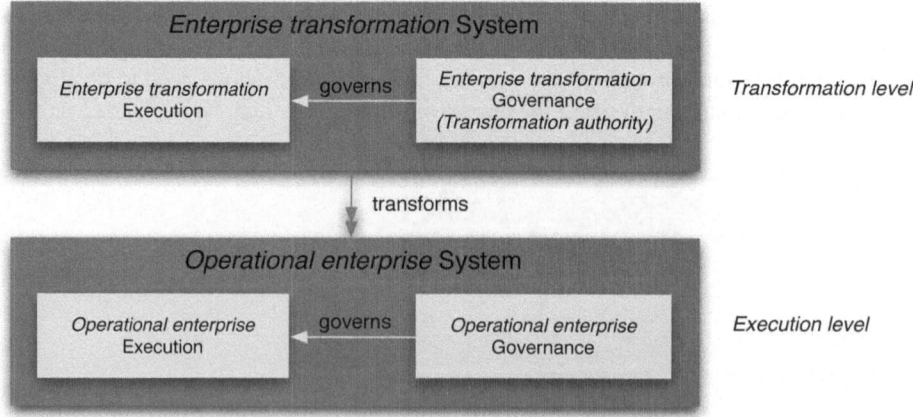

Fig. 3. Enterprise transformation governance

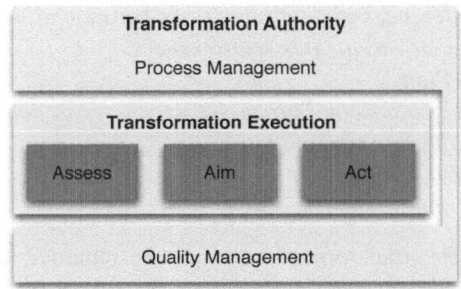

Fig. 4. Three perspectives on an enterprise transformation

output), and that this process is controlled by some (internal) authority (the *governing system*) which monitors, and if necessary adjusts, the process to make sure the intended objectives are reached. Since an organisation is part of a larger system, the governing system also interacts with the environment to determine which services of products to deliver, to determine new opportunities and to determine changes in the environment.

When combining the governance paradigm with the view on enterprise transformations as depicted in Figure 1, we arrive at the situation as depicted in Figure 3. Both at the execution level and the transformation level, a distinction is made between the processes which are the *target system* of the governing processes, and the governing system itself. The *enterprise transformation governance system* constitute the (operational part of) the *transformation authority*. A transformation authority involves two important (sub)systems:

Process management – The management of the processes involved in the execution of the actual transformation.

Quality management – Management of the quality of the results produced by the transformation processes.

while the actual execution of the transformation can be thought of as comprising three core processes:

Assess – The *assess*ment (diagnose) of the problem/challenge the transformation seeks to solve/meet.

Aim – The identification of how the transformation *aim*s to solve/meet the problem/ challenge (formulation/selection of the treatment).

Act – The *act*ing out of the actual transformation (performing the treatment).

This leads to the situation as depicted in Figure 4. Note: it is no accident that there are no arrows present. In general, the execution of the assess/aim/act processes will be highly iterative and cyclic in nature.

A comparrison can be made to Deming's [19] *Plan, Do, Check, Act* cycle for quality improvement. In terms of this cycle we would have the following mapping:

Assess – Involves Deming's notions of *Check*:
> *Measure the new processes and compare the results against the expected results to ascertain any differences*
> and *Act*

> *Analyze the differences to determine their cause, and determine where to apply changes that will include improvement.*

Aim – Corresponds to *Plan*:
> *Establish the objectives and processes necessary to deliver results in accordance with the expected output.*

Act – Corresponds to *Do*:
> *Implement the new processes.*

At an enterprise scale, transformations might be regarded at three key levels of granularity:

Project level – The level of specific projects having a clearly defined goal and time-frame in which to achieve this goal.

Program level – The level at which we consider several transformation projects contributing towards a larger overarching goal, still bound to a specific time-frame.

Portfolio level – The level at which enterprise transformation is regarded as a continuous collection of programs working towards the execution of the enterprise's strategy.

When taking these three levels into account, we end up with the situation as depicted in Figure 5. Note again: it is no accident that there are no arrows present. In general, the execution of the assess/aim/act processes will be highly iterative and cyclic in nature, and will even iterate and cycle between the levels.

4   Architecture as a Means for Informed Governance

Several socio-economical and technological trends drive enterprises to transform themselves. As discussed before, the diversity of these challenges and the resulting desires to transform (parts of) the enterprise, make it desirable to align all required and desired transformations. Judging whether a portfolio of enterprise transformations is well aligned, requires insight into the desired overall result as well as the planned and *achieved* effects of the individual transformations. It also puts a major challenge on an enterprise's management to make the right decisions at the right time and ensure that these decisions are translated into the right actions.

As discussed in [1], architecture offers a means for management to obtain insight, as well as to make decisions about, the direction of enterprise transformations. This is what we refer to as *informed governance*. During the *assess* and *aim* processes (see Figure 5), enterprise management needs insight into the way a transformation may improve the enterprise's ability to deal with these trends. Some concerns of enterprise management that need answers/insight:

- What is the rationale for this transformation? Will the transformation enable us to better deal/exploit with the socio-economic and technological trends?
- What are alternative transformation paths and their relative costs/benefits? What is the impact on current enterprise, its processes, structures, alliances, IT, et cetera? What are the risks during/after the transformation?
- What part of the enterprise will be impacted by the transformation?
- What are the relations and dependencies with other transformations/projects?

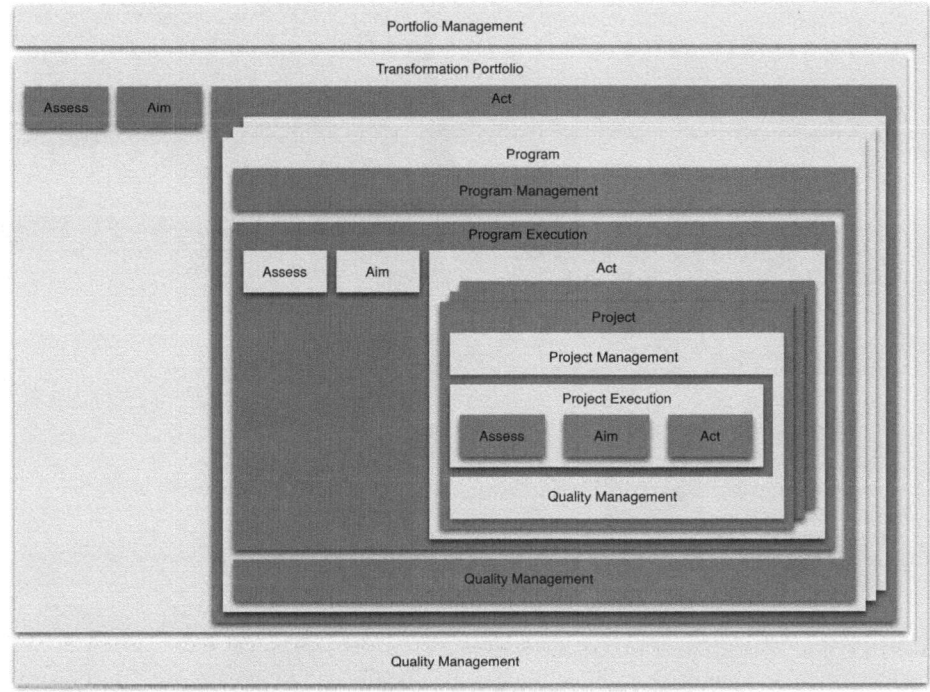

Fig. 5. Three levels of enterprise transformations

- When will the results of the transformation be effective?
- How sound is the business case for the transformation? What will it cost? How big are the benefits?
- What are consequences/opportunities of alliances with external parties or innovation networks?

More specific requirements on architecture as a means that follow from this (see [1]) are:

- Express/depict a coherent, comprehensive and concrete image of the desired future state(s) of the enterprise
- Provide a common language to a portfolio of changes/transformations of an enterprise.
- Identify a roadmap for the transformations needed.
- Distinguish between short-term solutions and long-term (structural) solutions
- Give a clear context and direction - limiting design freedom - to individual programs and projects that contribute to the desired overall transformation.
- Select available solutions and/or packages that are to remain or to become a part of the solution, whether in-house or sourced by a business partner.
- Enable traceability of design decisions from the strategic level via programs to specific projects.

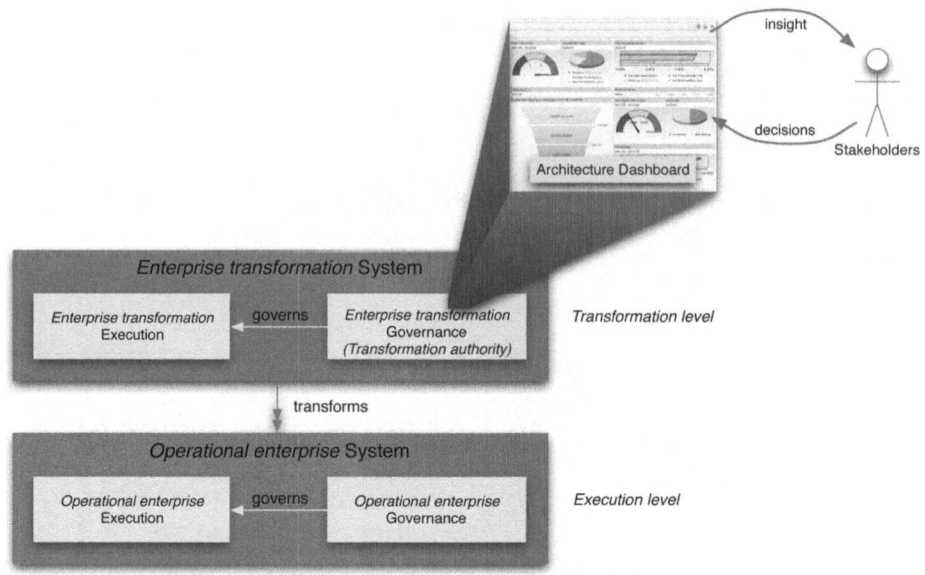

Fig. 6. Enterprise architecture on a dashboard

The use of an enterprise architecture can also be likened to the use of a "dashboard" which allows the architect and stakeholders to steer the enterprise's transformation processes. When using this metaphor, the *"dashboard"* displays the *architecture* in terms of relevant aspects of the current state of the enterprise, its future direction and the desired states of the enterprise.

Just as the selected/displayed speed, altitude and direction of an airplane is not *the dashboard*, but rather *displayed on* the dashboard, the dashboard is not the enterprise architecture. Analogously, it is the enterprise architecture, or rather a part thereof, what will be displayed on the dashboard. In addition, the dashboard may contain a report on the gaps between the current state and desired states, as well as its *operational performance* in terms of its current state. In an airplane, a *"dashboard"* may comprise of indicators (meters, lights, et cetera) and controls (levers, handles, pedals, and knobs). In the case of enterprise architecture as a means to govern transformations, the dashboard needs at least:

- *indicators* giving insight into:
    the enterprise's current state,
    the enterprise's current performance,
    the enterprise's future (expected) performance,
    the selected direction and progress of its transformation processes,
- *controls* allowing the transformation processes to be influenced:
    the desired state of the enterprise,
    plateaux of intermediary stages,
    overall regulations.

The indicators may take the form of models, views, performance measurements, et cetera. The controls may take the form of (enforced) reference models, design principles, standards, et cetera. This is illustrated in Figure 6. More specifically, architecture is used during the *assess* and *aim* processes to analyse problems in the current situation and formulate the desired target situation. In line with [20] the former would involve the use of a *base-line architecture*, while the latter leads to the *target architecture*. The *target architecture* serves as input to the quality management (sub-)process of the realisation as it will conducted by the *act* process.

In [1], seven key applications for architecture as a means have been identified. In combination, these applications provide an instrument to make informed decisions as well as to ensure compliance of the transformation to these decisions, at several levels of specificity:

Situation description – Use architecture as a means for goal/cause analysis to investigate problems/shortcomings in an existing situation. This also involves the creation of a shared understanding (among stakeholders) of the existing situation.

Strategic direction – Use architecture to express (and motivate) the future direction of an enterprise, as well as investigate (and evaluate) different alternatives. This also involves the creation of a shared (among stakeholders) conceptualisation of the (possible) future directions, and shared agreement for the selected alternative.

Gap analysis – Use architecture to identify key problems, challenges, issues, impediments, chances, threats, etcetera, as well as make well motivated design decisions that enable a move from the existing situation into the desired strategic direction.

Tactical planning – Use architecture to provide boundaries and identify plateaux (intermediary steps) for the transformation of the enterprise towards the articulated strategic direction. In this context, enterprise architecture is used as a planning tool, making the realisation of a strategy more tangible.

Operational planning – Use architecture to give a clear context and direction for a portfolio of projects working towards the realisation of the first plateau as defined at the tactical planning level.

Selection of partial solutions – Use architecture as a means to select one or more standard solutions and/or packages that are to become part of the solution and/or decide to outsource an entire business process/service to another enterprise.

Solution crafting – Use architecture to create the high level design of an actual step in the enterprise transformation as it will be realised (and implemented) in the context of a specific project.

When adding the architecture to Figure 5 we end up with the situation as depicted in Figure 7, where a distinction has been made between an *enterprise architecture* and a *solution architecture*. The enterprise architecture is concerned with a longer term regulative perspective giving directions to a number of transformation programs, while a solution architecture is more concerned with specific choices pertaining to transformation projects within a single transformation program. Needless to say that the solution architecture should comply to the over-arching enterprise architecture.

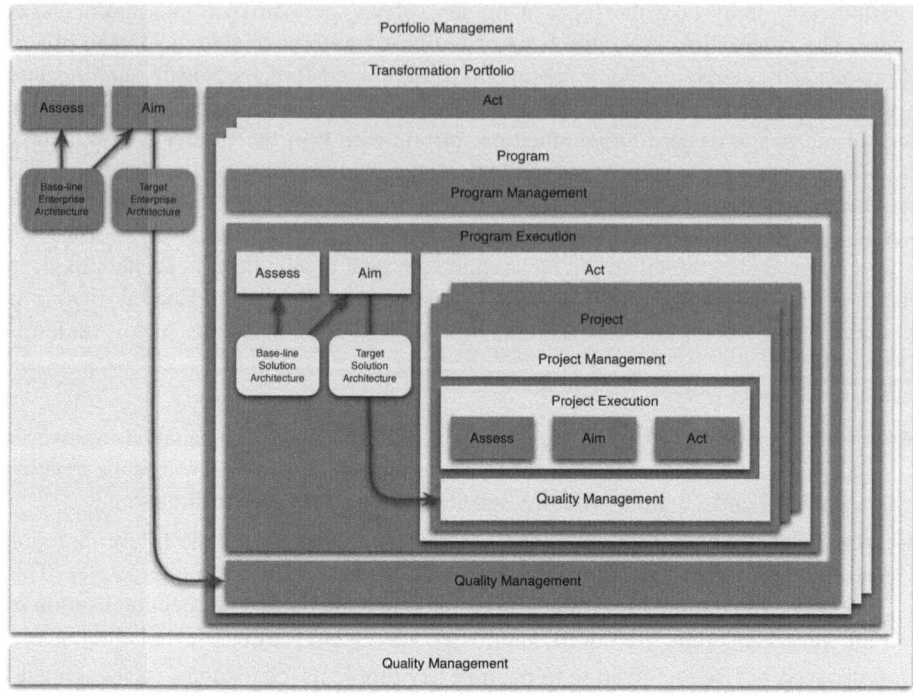

Fig. 7. The role of architecture

5   Mature Transformation Authorities

In this section we focus on the transformation authority as the function within an enterprise which is responsible for the activities involved in the governance of transformations. We start by identifying the requirements put on a *governing system* according to the governance paradigm (see Figure 2 and [2]):

- A governing system should have a goal with regards to the target system, providing guidance for steering.
- It should have information about the target system:
    - Its state (in the case of a moving object, this would include location, vector, speed and acceleration).
    - Environment variables influencing its state.
- It should have a (predictive) model of the behaviour of the target system, including its responses to steering signals.
    For example, to be effective, it needs an understanding of the current state (and evolution) of the enterprise, It also needs the ability to predict or gauge the effects of steering actions. Without this ability, the steering can only be reactive rather than pro-active.
- It should have enough requisite variety [4] to control the target system.

The latter requirement is one of the most challenging ones when implementing a transformation authority. First of all, it involves the challenge of involving senior

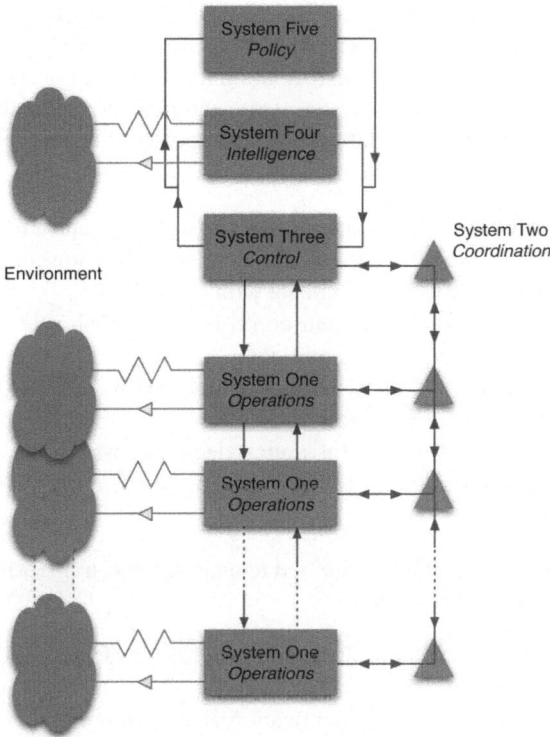

Fig. 8. Viable Systems Model

management of the enterprise in the decision making process leading up to an enterprise/solution architecture. This typically requires a collaborative approach aiming to create as much understanding and commitment as possible. This commitment is more than needed during the actual communication/enforcement of these decisions to the transformation execution. During the execution of a transformation one typically has the inclination to favour short-term interests over longer-term interests especially when time pressures mount. In such cases, the *process management* and *quality management* streams (see Figure 4) of transformation governance are likely to clash as well. The decisions made in relation to the architecture tend to focus on the longer term interests/qualities of the portfolio of current/future enterprise transformations, while short-term pressures will favour faster and/or cheaper realisation from the perspective of a single project or program. These shorter-term versus longer-term clashes require commitment from senior management towards the original decisions. More details on strategies to achieve committed to decisions by senior management can be found in e.g. [21,22,23].

In addition to the trade-offs between longer-term and short-term interests, the directions formulated at the portfolio and/or program level (see Figure 7) may not always provide enough guidance to specific projects and/or may not be workable in a given practical situation. This also implies that a transformation authority should be willing to engage in the actual execution of the transformation, and be willing to learn how to make their directions (e.g. the architecture) more useful from a project/program's perspective.

More insight into the role of a transformation authority can be achieved by considering an enterprise transformation system from the perspective of the viable systems model [5]. The viable systems model takes the perspective that an organisation, such as an enterprise transformation system, comprises five key sub-systems:

System one (operations) – The primary activities of the organisation.

System two (co-ordination) – Communication & co-ordination channels allowing the activities of system one to communicate with each other.

System three (control) – Structures and controls to establish rules, resources, rights and responsibilities of system one and two.

System four (intelligence) – Looks outward to the environment to monitor how the whole system needs to adopt to remain viable.

System ve (policy) – Responsible for policy decisions within the organisation as a whole to balance demands from different parts of the organisation, and steer the organisation as a whole.

This is illustrated in Figure 8. When applied to an enterprise transformation system, we would have the following five systems:

System one (operations) – The projects and programs executing the enterprise transformation.

System two (co-ordination) – Communication & co-ordination between transformation programs and projects.

System three (control) – Structures and controls to establish governance of enterprise transformations, including the formulate and deployment of enterprise architectures.

System four (intelligence) – Monitor the environment for developments which require changes in the enterprise transformation system and/or the architectures used.

System ve (policy) – Strategic management of the transformation authority. Monitors the effective functioning of the authority, making changes where/when needed in terms of structures and policies.

where, in terms of Figure 2 systems one and two are the *target system* and systems three, four and five are part of the *governing system*, i.e., the *transformation authority*.

Systems four and five are actually "new" in our discussions. As suggested by Figure 3, the core of the transformation authority's activities are formed by *enterprise transformation governance*, i.e., system three. System four and five essentially require a transformation authority to be able to change and improve itself, i.e., mature itself. This brings us to the introduction of the concept *maturation* of a transformation authority, the key driver of this chapter. In terms of Figure 3, this leads to the introduction of a *maturation level* on top of the *transformation* and *execution* levels. The result is illustrated in Figure 9.

An important aspect in the maturation of a transformation authority is the evaluation of its effectiveness. This effectiveness may differ from one organisation to another. In [24] a model is described on how to measure the effectiveness of an organisation's

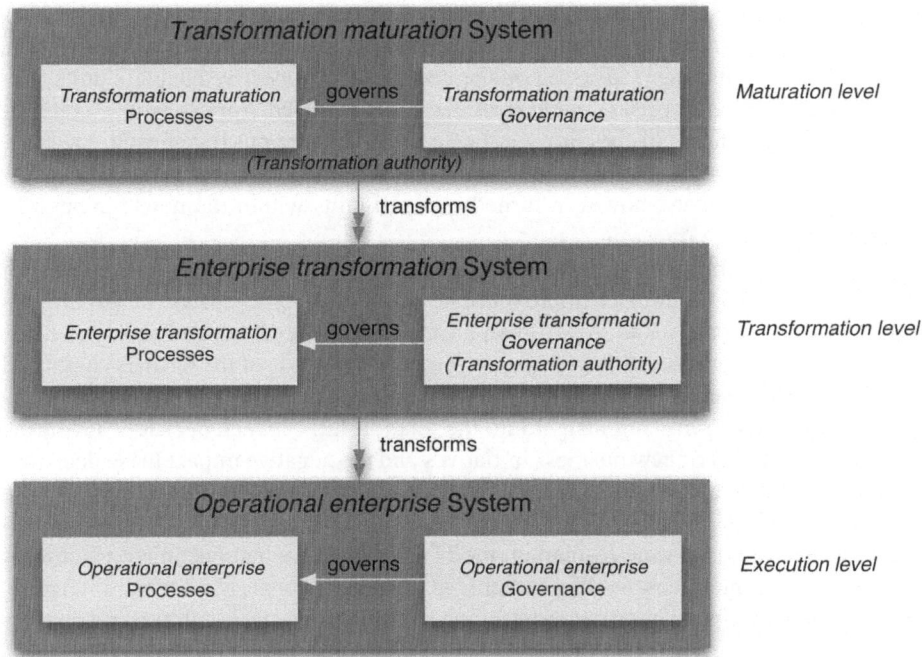

Fig. 9. Adding a maturation level

architecture function. An architecture function as defined in [24] is close to our concept of a transformation authority. However, we prefer to stress the primary role of the transformation authority, being the governance of transformations, rather than to refer to one of the *means* (albeit an important one) it uses to do so: *architecture.* An important indicator for the effectiveness of a transformation authority is the maturity level at which it operates. Analogously to the capability and maturity model for software development processes, maturity models have been developed to express an organisation's maturity in using architecture. A prime example is the USA Department of Commerce's maturity model [25].

Note that any enterprise executing enterprise transformations does have a transformation authority at some level of maturity. This might be at a purely "ad hoc" and totally immature level. The "implementation" of a transformation authority therefore can be seen as taking the enterprise's transformation authority to at least the first maturity level.

6   Case 1 – Post-Merger Improvement of a Transformation Authority

6.1   Situation

A large Netherlands-based multinational enterprise highly depends on IT for its core business processes. Over the past fifteen years, this organisation has grown to its current size as a result of several mergers and acquisitions. About five years ago, the organisation found itself hampered by a lack of synergy between the different business units

originating from past acquisitions and mergers. To a larger extent this was traced back
to the lack of integration between the IT departments of the different business units,
as these departments where focussed on the needs of the respective business units (formally independent enterprises) rather than the needs at the enterprise level. To remedy
this, the IT departments of the different business units were put together in one large
IT business unit responsible for all IT operations for the entire enterprise. The original IT departments were, however, turned into sub-units within the newly created IT
department.

Two years ago, much to the disappointment of top management, it was concluded
that the situation had not really improved. The sub-units within the new IT department
still maintained strong ties to the original business units they belonged to. As a result,
during system development/transformation projects, the needs of the business units still
prevailed over the needs of the enterprise as a whole.

Even more, due to missing insight into the relationship between decisions favouring
a fast time-to-market of new business initiatives and the negative impact these decisions
may have on the maintainability of the IT infrastructures, most key decisions had been
made based on the shorter term interests. Meanwhile as a result of ill-guided pragmatism in achieving fast a time-to-market, the IT infrastructure had become a patchwork
of platforms and interfaces, leading to a choice high operational risks for the enterprise.
Recently, this hidden danger surfaced, leading to disruptions of the enterprise's primary
processes.

6.2   The Improvement Programme

To improve on the existing situation, an improvement programme was put in place
by senior management of enterprise, aiming to improve transformation governance of
the IT department. The focus of this transformation programme was on: increasing the
maturity of transformation processes, training of people in architectural standards, the
standardisation of the development/transformation processes, as well as the creation of
an explicit enterprise architecture.

Management of the enterprise, and senior management of the IT department, adhered
to a rather Anglo-Saxon management style. As a result, the improvement programme
used a top-down approach to initiate the changes needed. However, most of the lower
management, as well as the people involved in IT operations and development projects
where brought up in a more consensus based culture. This obviously produced a lot of
tension within the organisation.

However, the strong top-down approach did enforce commitment from middle-level
management during the transformation. When using a bottom-up approach based on the
consensus culture, while creating intrinsic motivation by increasing awareness/insight
into the advantages of a more mature way of working, the improvement programme
might still have failed due to a lack of commitment from middle-level management. In
the end, the combination of using a strict top-down approach forcing management within
the IT department to move into the right direction, and increasing the knowledge level of
people working in development/transformation projects by having them attend courses,
proved to be a balanced approach. The top-down approach also brought about a kind of
a "shock-therapy" in the sense that a lot of the scepticism about the IT department's own

ability to create standards and workable architectures was swept away when the initial results were produced. The forced adoption of these standards consequently led to focussed discussions about the correctness of the standard. However, rather than paralysing progress in creating consensus based standards, the discussions were now more easy to focus on the improvement and fine-tuning of the top-down created standards.

While the improvement process took hold, it became clear that it was necessary to also increase the maturity of senior management in the IT department as well as the awareness of the impact of design decisions with business management. Thus far it was highly difficult to clearly trace back increased maintenance costs to undisciplined business desires. The next logical step was therefore to create reference architectures for standard solutions with low maintenance costs, offering a clear choice to business management: standardise or be prepared to pay extra. As a result of increased training of the people in the IT department, their personal esteem also increased, enabling them to indeed more clearly and *confidently* communicate these trade-offs back to business management.

During the improvement process, a complicating factor was that some people involved played roles in both the improvement programme and the daily workings of the IT department. The sponsors of the maturation programme needed maturing themselves as well in their role in the IT department, in particular in their senior management role.

6.3 Conclusion

In the first attempt by the enterprise to re-organise the existing IT departments, not enough attention was paid to integrating the former departments into a new whole, including the setting up of new relations (and ways of "doing business") between the new IT department and the business units, in particular the establishment of an able and effective transformation authority.

In the second attempt, a very interesting balance was struck between a top-down approach, creating a "shock-therapy" effect, and the enabling of people by training, also balancing between a top-down management style with a consensus based culture. A crucial factor also turned out to be the ability of the sponsors of the improvement process to realise that they themselves needed to mature as well with regards to their role in the transformation governance processes.

The top-down enforced creation of standards and reference architectures also emphasises the importance, in this case, of balancing the execution of a transformation authority's tasks with the progress of its maturation process.

7 Case 2 – Orchestrated Improvement Programme

7.1 Situation

A large Netherlands-based international financial institution (34000 fte), in an attempt to better align business with IT, is looking for ways to improve the governance of the information systems development process, encompassing a fairly large IT organisation of about 2000 fte in total. The institution wants to achieve better alignment between the business needs and the process changes and IT solutions delivered by the individual projects (about 500 projects are in portfolio). Earlier, relatively isolated, attempts focusing on improvement of the portfolio management process, development of a software factory and development of a corporate IT architecture failed or were very ineffective. The institution is facing a dilemma: to spend even more money "to get things right" in an integral manner or muddling on using the results of earlier improvement initiatives. After much deliberation, the choice is to start a *software process improvement program*. The overall objective of this programme is the alignment of business and IT. Sub-goals include development productivity improvement and increasing (perceived) quality, speed and professionalism. The productivity improvement turns out to be the main management trigger, as it enables a significant cost reduction while improving the quality of the project results.

7.2   The Improvement Program

The software process improvement programme consists of several change projects: introduction of an iterative enterprise transformation method, the implementation of a CMM level 3 compliant quality system, the introduction of a measurement dashboard and associated measuring instruments, special attention to professional attitude and, moreover, *harvesting* products of earlier initiatives. One of these initiatives is the top-down development of a large scale enterprise architecture. This architecture consists of a large number of guidelines, principles, rules, standards and reference models meant to guide and jump-start projects. These initiatives failed, largely because of the disconnection between the theory "invented" by the architects and the real need of the projects. However, a large body of knowledge was delivered.

The software process improvement programme uses a combination of top-down and bottom-up to implement the results of the various projects and initiatives. Bottom-up, because individual projects are used to gradually introduce the results. Top-down, because the boundaries of the implementation are very clear to all stakeholders, with a very high commitment of senior management wanting to achieve the programme goals. The combination proves to be very successful. In a time span of 3 years, the entire development organisation is transformed into a CMM level 3 organisation, using one, user-centred iterative development method, measuring its processes and their outputs and, for the first time ever, real *control* over the development project portfolio. This results in a good alignment between business and IT on both strategic (portfolio), tactical (project) and operational level.

The enterprise architecture is gradually introduced in the projects. A relatively small group of enterprise architects, headed by a Chief Architect, is formed, aiming to support the individual projects as much as possible using the components of the enterprise architecture. These architects act as consultants, paid by the software process improvement program. Once the individual projects and programs, and more importantly their *business owners*, see the added value of the enterprise architecture and the architects, the management of the financial institution decides to make those parts of the enterprise architecture compulsory that have been successfully applied in projects and programs. In this decision, the chief architect acts as a consultant. Over the course of the three years, large parts of the enterprise architecture are introduced in this manner.

7.3 Conclusion

Especially the gradual introduction of enterprise architecture in the context of the larger software process improvement programme proves to be successful. Projects and programs are not confronted with large piles of documents, but only with those parts of the Enterprise Architecture that really guide and help them. A relatively large period of time (3 years) proves also to be of importance: things have to "sink in" and much persistence is needed for that. Another critical success factor is the attitude of the architect: a supportive consultant, not an arrogant know-it-all. Harvesting is the fourth success factor: re-use existing products instead of inventing the wheel again. Last but not least is the level of alignment of introducing an enterprise architecture with other initiatives to improve governance & projects such as the implementation of the enterprise transformation method, as well as the incorporation of many related initiatives and projects into one single programme.

The big challenge in these initiatives is sustainability. During the three years of the program, management attention is quite high. Shortly after the end of the program, the financial institution is involved in a new change initiative, which redirects management attention. Sadly, this impairs most results in quite a substantial way.

8 Analysis of the Cases

Obviously, the failure or success of a maturation/implementation of a transformation authority is largely determined by the characteristics of its execution process and the characteristics of the process aiming to mature this execution process, including the actors, products, et cetera involved. Moreover, the interaction between execution of the transformation authority's processes and the progress of the maturation process is an important determinant for success, as stressed in the case 1. The road to success would then be, to create execution and maturation processes with the 'ideal' characteristics. This, however, poses some problems:

- It is not possible to define 'ideal', as each situation is different and constantly changing;
- Suppose it would be possible to define 'ideal', it is hard to assign the 'right' values to characteristics.

A more practical way of looking at this, is to consider the *road blocks* for successful implementation of a transformation authority. Success is, in this view, achieved by *removing* the roadblocks. In the cases, several roadblocks were considered:

- The inconsistency between management culture and the culture of the 'shop floor'. In case 1, the management culture is very Anglo-Saxon and directive, whereas the culture of the rest of the organisation is more of a consensus-based nature. These cultures clash, especially in the case of transformation and its governance process;
- The lack of real management attention and support. In case 2, it turns out that, after the software process improvement program has ended, and management redirects its commitment to other initiatives, earlier achievements of the maturation program are eroded. It seems that the organisation has forgotten how well transformation governance went when management had the right attention;

- The inconsistency in communication, not "walking the talk". Management in case 1 used only one directive communication style, but the directives were not or hardly implemented. People were not informed about the rationale of certain management decisions, resulting in mistrust and vague implementation processes. Another related, roadblock was the lack of setting the good example by management: "practice what you preach". Management made an exception for itself not to follow the directives that were applicable to the entire organisation;
- The inability of architects to show results and add real or perceived value. The social and communicative competencies of the architect and other important stakeholders in the transformation governance process are of utmost importance. In case 2, the architects that designed the instruments to improve the systems development process learned how to act as a consultant, helping the projects that were using the development process. In case 1, architects used a more traditional ivory tower approach with resistance from the actors that were subject to the transformation as a consequence;
- The inability of actors in the transformation to play several roles, connected to different interests. In case 1, several actors, especially important ones like the sponsor, played roles in both the operational processes, the projects and programs changing those operational processes *and* the program to implement/mature a transformation authority. It requires special skills to be able to separate the interests related to the different roles, especially with respect to prioritisation and the correct behaviour dependent on the role one plays;
- The ineffective way signals from the shop floor are received by management. In case 1, they were more or less treated as threats, whereas in case 2 management did not want to listen anymore after the program ended. The workers in the projects that were subject to change often know best how to improve the way of working, and when these signals are not picked up, potential improvements are lost;
- The political forces and hidden agendas in an organisation. Often, the issue is dealt with in a rational way, whereas quite a number of irrational factors play a much more important role. In both cases, politics and irrationality were roadblocks.

Notice that roadblocks are typically *not* to be found in the artefacts of the various transformation initiatives, but in their context. Their removal is part of the change management required to implement the artifacts.

9   Transformation Maturity Framework

Based on the analysis as discussed in the previous section, this section discusses a broad maturity framework for the maturation of transformation authorities, taking several important aspects into consideration that may lead to blockages, and maybe even lead to erosion of already achieved results.

9.1   Implementation Strategies and Situational Factors

The cases show some best and worst practices regarding the implementation of an transformation authority. Some strategies work, some do not. In general, we have seen that the success of an implementation is determined by:

- The characteristics of the implementation process;
- The characteristics of the situation in which the transformation authority is implemented.

The process of implementing/maturing an transformation authority is not very different from processes aiming at organisational changes (see, for instance [26]) in general. These processes boil down, to put it very simplistically, to a combination of top-down vision setting, initiating, managing, communicating, measuring and supporting versus the bottom-up ability to execute, competencies, willingness, enthusiasm, learning and showing success. Irrespective of using a top-down or bottom-up approach when introducing an transformation authority, the awareness of the importance of architecture as a means to steer developments, needs to grow – it is not viable to implement a full-fledged transformation authority right from the start. It will need time to learn and to be able to support the enterprise transformation projects and programs in the best possible way. We have seen that architecture always needs to balance longer-term interests to short-term interests. In this present day and age and faced with day-to-day pressures, one tends to forget about the longer term interests. Especially management is very focused on achieving short-term results, which poses a huge risk on the potential success of a transformation authority.

Implementing/maturing an transformation authority therefore requires a deliberate strategy in which the key players are made aware of the interplay between short-term and long-term interests, while gatekeepers are introduced to safeguard the long-term interests against erosion by short-term considerations. To be able to develop such a deliberate strategy a more explicit understanding of the role of architecture as a means of steering is needed, as well as the underlying reasoning. We have seen in the cases that this role is very much dependent on situational factors, such as:

- The competencies of the architects, especially with regard to communication and other personal skills;
- The historical perspective of earlier attempts to implement an transformation authority;
- The focus, attention and support of the management;
- The extent to which a "burning platform" exists: is it really necessary to implement a transformation authority?
- The extent to which the architecture implementation is connected to other improvement initiatives;
- The culture of the organisation and its parts, especially management culture versus "shop floor" culture;
- The maturity of project and program management;
- The scope, depth and size of the architecture itself.

There is obviously no silver bullet for successful implementation of a transformation authority. One conclusion we can draw from the cases is, however, that the "maturity" of the organisation plays a crucial role. We will elaborate this in the next sections.

9.2 Removing Roadblocks: The Transformation Maturity Concept

In both cases we found that the extent to which roadblocks can be removed is dependent on a concept we call "transformation maturity". We define transformation maturity as the capability to achieve effective governance of transformation processes, where "effective" is obviously situation-dependent. We use the number of roadblocks as a proxy to the transformation maturity level: the higher the number of roadblocks, the lower the transformation maturity level. A high number of roadblocks is a symptom of low transformation maturity, and this can hamper the transformation significantly.

For the time being, we consider transformation maturity as a relative notion, so we can only *compare* the transformation maturity of two distinct systems. To enable comparison and provide a more generic analysis of the case, we use the levels identified in Figure 9: *execution level*, *transformation level* and *maturation level* in relation to enterprise transformtions. At each of these levels, we can refer to transformation maturity:

1. On the execution level, we consider the *operational enterprise* system. In this system, the operational processes of an enterprise are executed and governed. This system encompasses the execution and governance of primary and secondary processes. In case 2, for example, issuing loans or trading stocks.

   At this level, transformation maturity is related to the system's ability to undergo transformations.

2. On the transformation level, we consider the *enterprise transformation* system. This is the system that transforms the *operational enterprise* system. It consists of transformation processes and the governance of those processes. Typically, these processes are information system implementation projects and programs, for instance a large ERP implementation or a custom development project.

   At this level, transformation maturity refers to the quality and effectiveness of the transformation execution and the activities of the transformation authority.

3. On the maturation level, the *transformation maturation* system is considered. This is the system that aims to transform the *enterprise transformation* system, in order to improve its transformation maturity. Again, the system consist of the actual maturation processes and the governance of these processes. Examples of maturation processes are the software process improvement program in case 2 and the project to implement the transformation authority in case 1.

   At this level, transformation maturity is concerned with the ability of the *transformation maturation* system to continuously improve the maturity of the *enterprise transformation* system.

The transformation maturity concept and the distinction of three systems in the transformed enterprise enable us to analyse the cases in a different way.

In case 1, the transformation maturity of the *operational enterprise* system is at about the same level as the transformation maturity of the *enterprise transformation* system: the number of roadblocks is more or less comparable in both systems. The transformation maturity of the *operational enterprise* system is improved by the *enterprise transformation* system. However, the transformation maturity of the *transformation maturity* system is lower than the other two systems, because there are (relatively) much more roadblocks in this system. Due to this, the *transformation maturity* system is hardly able to transform the *enterprise transformation* system.

In case 2, there are two situations: the states during and after the software process improvement program.

During the Program, the *transformation maturity* system has a relative high transformation maturity level compared to the *enterprise transformation* system. A lot of roadblocks in the Program are not present or have been removed. The Program is quite able to transform the systems development and project management processes, i.e., the *enterprise transformation* system. The latter, on turn, becomes, as a result of the Software Process Improvement Program, more mature than the *operational enterprise* system.

This changes after the end of the Software Process Improvement Program. A roadblock is introduced in the governance of the *enterprise transformation* system: lack of management attention. Because the *operational enterprise* system has matured due to the improved transformation ability of the *enterprise transformation* system as a result of the Software Process Improvement Program, the relative transformation maturity of the Enterprise transformation Program is decreasing as compared to the *operational enterprise* system. This, on turn, decreases transformation ability of the Enterprise transformation Program, resulting in lower project productivity and lower alignment between business and IT. This is exactly what happened in case 2. In a comparable way, the transformation maturity of the *transformation maturity* system also decreased, as a consequence of the addition of some roadblocks on this level.

9.3   The Transformation Maturity Framework

The case analysis in relation to the introduction of levels in Transformation Governance and the Transformation Maturity concept has provided us with the following important condition for successful transformation governance: System A, transforming System B, must have at least the same transformation maturity as System B. Applied to the three relevant systems we identified in a semi-formal notation (where TM stands for Transformation Maturity):

$$TM(\text{Transformation maturation System})$$
$$\geq TM(\textit{enterprise transformation } \text{system})$$
$$\geq TM(\textit{operational enterprise } \text{system})$$

This is principle 1 of our *transformation maturity framework* (TMF). Enterprise transformations complying with this principle are called "ideal".

As a consequence, it does not make much sense (or, at least, it is ineffective) to transform processes with a relative immature system. Before starting a transformation, roadblocks in the transforming system have to be removed to ensure a sufficient transformation maturity level and "ideal" enterprise transformations.

A second observation is, that transformations are always an interplay of bottom-up and top-down actions. Typically, the "pain" comes from below, i.e., the systems that are transformed. "Healing" this pain is supposed to come from above, the transforming system. Healing consists of, for example, providing an artefact or removing a roadblock. If the cure is structural, it can be considered a transformation. In non-ideal enterprise transformations, sometimes a lot of pain is generated and in reaction, the transforming system tries to heal this pain in an ad-hoc fashion. As we have seen in case 1, there is often a trade-off between pain and healing. In much the same manner, a trade-off exists between preventing and healing: preventing being a pro-active activity and healing as reaction to pain. In an ideal transformation governance, everything can be foreseen and thus prevented - but reality is different, so there is still healing to be done.

The second principle of the TMF is that in enterprise transformations, there is always a trade-off between opposite factors, often in terms of healing vs. pain and healing vs. preventing.

The second principle implies that an enterprise transformation is an iterative process, aiming at an "ideal" situation with the right transformation maturity levels and constantly seeking the trade-offs between healing and pain and between preventing and healing. In the cases we observed that it does not make much sense to produce all the artefacts in advance, because reality is changing constantly and artefacts have to connect to this reality instead of being theoretical. On the other hand, it is not possible to start completely without artefacts, so a certain minimal set has to be there before the journey to the ideal situation can start.

The third observation relates to the actors in the enterprise transformation. As mentioned in the list of roadblocks, many actors play different roles. If these roles are played on different levels, the actor involved has to be aware of this. He has to be able to deal with the different interests and the different objectives and priorities in systems on different levels. For example, a sponsor is often a line manager on a fairly high organisational level. As a line manager, he might have other interests, priorities and objectives than as a sponsor. If he cannot separate the various roles he is playing, the number of roadblocks can increase. This is not necessarily bad, as long as it occurs on the 'lower' levels, notably the Execution level. If it occurs on higher levels, for instance the Maturation level, the consequence is a lower maturity and a lower chance of achieving the "ideal" situation.

We translate the observation in the third principle of the TMF: An actor has to be able to play multi-level roles.

10   Conclusion

The driving interest of this chapter was the implementation of mature *transformation authorities*. To this end, we took both a theoretical and a practical perspective on enterprise transformation and their governance.

We started by defining more precisely what we mean be *enterprise transformations* and their *governance*, as well as the need to use *architecture* to achieve *informed governance*. The concept of a *transformation authority* was introduced as the function in an organisation which is responsible for the governance of enterprise transformations.

We then continued with a discussion of two cases drawn from our industrial practices. Using the insight from these cases, we then refined our theoretical considerations in terms of the *transformation maturity framework* (TMF).

As a next step we intend to further refine our definitions on the basis of practical experiences. More specifically, we aim to further refine the maturity model in terms of situational factors (such as the one discussed in Section 9) influencing the aptness for different strategies to mature/implement transformation authorities, as well as concrete roadmaps to indeed grow/mature transformation authorities.

References

1. Op't Land, M., Proper, H., Waage, M., Cloo, J., Steghuis, C.: Enterprise Architecture – Creating Value by Informed Governance. Springer, Berlin (2008),
   http://www.springerlink.com/content/978-3-540-85231-5
2. Leeuw, A.d.: Organisaties: Management, Analyse, Ontwikkeling en Verandering, een systeem visie. van Gorcum, Assen, The Netherlands, EU (1982) (in Dutch)
3. Leeuw, A.d., Volberda, H.: On the Concept of Flexibility: A Dual Control Perspective. Omega, International Journal of Management science 24, 121–139 (1996)
4. Ashby, W.: An Introduction to Cybernetics. Chapman & Hall, London (1956)
5. Beer, S.: Diagnosing the System for Organizations. Wiley, New York (1985)
6. Hagel III, J., Armstrong, A.: Net Gain – Expanding markets through virtual communities. Harvard Business School Press, Boston (1997)
7. Horan, T.: Digital Places – Building our city of bits. The Urban Land Institute (ULI), Washington DC, USA (2000) ISBN-10: 0874208459
8. Mulholland, A., Thomas, C., Kurchina, P., Woods, D.: Mashup Corporations - The End of Business as Usual. Evolved Technologist Press, New York (2006)
9. Tapscott, D.: Digital Economy – Promise and peril in the age of networked intelligence. McGraw–Hill, New York (1966)
10. Government of the USA: Sarbanes-Oxley Act of 2002. Number H.R.3763 (2002)
11. Porter, M.: What is strategy. Harvard Business Review (1996)
12. Treacy, M., Wiersema, F.: The Discipline of Market Leaders – Choose your customers, narrow your focus, dominate your market. Addison Wesley, Reading (1997)
13. Graaf, E.v.d.: Architectuurprincipes en clustercriteria voor de afbakening van out-sourcebare kavels. Master's thesis, Radboud University Nijmegen (2006) (in Dutch)
14. Op't Land, M.: Towards Evidence Based Splitting of Organizations. In: Ralyté, J., Brinkkemper, S., Henderson-Sellers, B. (eds.) Proceedings of the IFIP TC8 / WG8.1 Working Conference on Situational Method Engineering: Fundamentals and Experiences (ME 2007), Geneva, Switzerland. IFIP Series, vol. 244, pp. 328–342. Springer, Berlin (2007)
15. Op't Land, M., Dietz, J.: Enterprise ontology based splitting and contracting of organizations. In: Proceedings of the 23rd Annual ACM Symposium on Applied Computing (SAC 2008), Fortaleza, Ceará, Brazil (2008)
16. Rijsenbrij, D.: Outsourcing zonder enterprise architectuur lijkt op autorijden zonder veiligheidsgordel. Technical Report NIII–R0404, Nijmegen Institute for Information and Computing Sciences, University of Nijmegen, Nijmegen, The Netherlands, EU (2004)
17. Oxford University: Oxford Dictionary of English. Oxford University Press, Oxford (2005) ISBN-13: 9780198610571
18. Bunge, M.: A World of Systems. Treatise on Basic Philosophy, vol. 4. D. Reidel Publishing Company, Dordrecht (1979)
19. Deming, W.: Out of the Crisis. MIT Center for Advanced Engineering Study, Massachusetts, USA (1986) ISBN 0911379010
20. The Open Group: The Open Group Architecture Framework (TOGAF) Version 8.1.1, Enterprise Edition (2007), http://www.togaf.org
21. Nabukenya, J.: Collaboration Engineering for Policy Making: A Theory of Good Policy in a Collaborative Action. In: Proceedings of the 15th European Conference on Information Systems, pp. 54–61 (2005)
22. Nabukenya, J., Bommel, P.v., Proper, H.: A theory–driven design approach to collaborative policy making processes. In: Proceedings of the 42nd Hawaii International Conference on System Sciences (HICSS-42), Hawaii, USA. IEEE Computer Society Press, Los Alamitos (2009)

23. Nakakawa, A., Bommel, P.v., Proper, H.: Quality Enhancemenets in Creating Enterprise Architecture – Relevance of Academic Models in Practice. In: Practice-driven Research on Enterprise Transformation (submitted, 2009)
24. Raadt, B.v.d.: Normalized Architecture Organization Maturity Index. Technical report, Capgemini (2006)
25. Department of Commerce, Government of the USA: Introduction - IT Architecture Capability Maturity Model. Government of the United States of America (2003),
    `http://ocio.os.doc.gov/groups/public/doc/os/ocio/oitpp/`
    `documents/content/prod01_02340.pdf`
26. Senge, P., Kleiner, A., Roberts, C., Roth, G., Ross, R., Smith, B.: The Dance of Change – The Challenges to Sustaining Momentum in Learning Organizations. Doubleday, New York (1999)

Strategy and Architecture – Reconciling Worldviews

Bas van Gils

Strategy Academy
Parklaan 1, 3016 BA, Rotterdam, the Netherlands
b.vangils@strategy-academy.org

Abstract. The relationship between strategy and enterprise architecture is troublesome in many organizations. It seems that this cumbersome relationship is
similar to the more 'traditional' tension that seemingly exists between business
and IT. This paper explores three underlying causes of this tension, most notably (1) overlap in domain of expertise, (2) different languages and (3) different
underlying worldviews. It is argued that there is no single solution to resolving
this tension. Instead, the tension should be seen as a polarity that must be managed continuously. Only by ensuring that both groups of practitioners have a
shared understanding of the issues that the firm faces and are committed to resolving them together can the tension between these groups be relieved.

1 Introduction

Business and IT have had a long and troublesome relationship ever since computing
entered the business realm. The initial promise of IT was to make life easier by automating repetitive tasks, as well as to improve speed and accuracy. Living up to this
promise has turned out to be difficult to say the least. Initially this may have been
caused by the instability of computing machinery. However, as these machines became more and more stable it turned out that the complexity of large software systems
was the biggest issue. These days, it appears that the complexity lies in managing a
portfolio of many such systems.

Since that time a multitude of articles has been published about that cumbersome
relationship by professionals and academics, focusing on concepts and philosophies to
solve the issues. This includes purely technical solutions such as component based
development, object orientation and SOA [1,2,3] to management approaches to solve
the issues at hand. One of the first approaches to dealing with the issue of aligning
business and IT, both at the strategic level and tactical/operational level, goes back to
the late 1980s [4,5]. The fact that this issue still is highly relevant in many organizations is illustrated by the following example:

Example 1 - In the Netherlands, there are 25 autonomous (local) police forces as well
as one national force (the KLPD). The police make heavy use of IT systems to support
their work. In the 1980s and 1990s it became apparent that synergies could be gained
by organizing IT support centrally. As a result, around 2000, a "demand organization"
(responsible for finding out what the police forces need, and translating this into

E. Proper, F. Harmsen, and J.L.G. Dietz (Eds.): PRET 2009, LNBIP 28, pp. 181–196, 2009.
© Springer-Verlag Berlin Heidelberg 2009

projects) and "supply organization" (responsible for developing and maintaining systems) were created. Several years of mixed successes later, these two organizations merged into a single organization (vtsPN) which has the mission of "to help make the Netherlands safe". Even in the new situation, a tension between flexibility of the local police forces on the one hand, and fixing IT systems on the other remains.   □

A more recent approach to aligning business and IT is the concept of Enterprise Architecture (EA) [6,7,8,910,11]. The field of Enterprise Architecture emerged in the mid 1990s and has led to a series of publications, active communities with both practitioners and academics, and conferences (e.g., NAF[1] and LAC[2] in the Netherlands). The necessity of seeing the firm as a holistic entity in which business and IT must be managed integrally is an important aspect of the DNA of enterprise architects.

Unfortunately though, the EA community is still struggling in living up to this promise which was exemplified at the 2008 edition of the LAC conference. In the first keynote by Daan Rijsenbrij it was pointed out that enterprise architects still have the image of being 'only about IT' which could be one of the reasons why EA as a discipline is not always regarded highly by 'outsiders'. Secondly, the conference book mentions in its preface that it is remarkable that the topic of business/IT alignment has had a negative connotation for years, and that the time is ripe to study the fields of enterprise governance, enterprise engineering and enterprise architecture from a broader perspective; not only from the point of view of IT, but also from a business point of view [12].

In line with these observations, the general consensus at LAC 2008 seemed to be that the ties between enterprise architects and strategists (i.e., the boardroom level) should be strengthened, especially in tracks such as *Enterprise Architecture: strategic specialism for informed governance"[3]*. In acknowledgement of this fact, the Open Group has articulated that effective management and exploitation of information through IT is the key to business success (competitive advantage), and that a good enterprise architecture enables firms to achieve the right balance between IT efficiency and business innovation [11]. Unfortunately however, the evidence in the form of large-scale quantitative research, to support this statement seems to be lacking still.

In my view, the tension between business and IT has been transposed to a tension between strategists and enterprise architects as these groups seem to be most involved with this issue. In this paper I will explore the underlying issues that cause these tensions and make a suggestion on how to deal with them in day to day practice. However, the following section starts with a brief survey of literature related to strategic management on the one hand, and enterprise architecture on the other.

2 Definitions

This section provides a brief overview of current literature on enterprise architecture and strategic management as the groundwork for the discussions to come.

---

[1] http://www.naf.nl
[2] http://www.lac2008.nl
[3] The Dutch name of the track is: Enterprise architectuur: strategisch specialisme voor inhoudelijke sturing.

2.1 Enterprise Architecture

As mentioned in Section 1, the field of Enterprise Architecture emerged in the 1990s as a "successor" of the business/IT alignment initiatives of the 1980s. Over the last few years a heated debate has taken place on the question of what EA *is*. Indeed, a wide variety of definitions, frameworks and approaches have been developed and it seems that the community has reached the point where pragmatism (getting things done) takes precedence over definitions. The following quote by Jan Bosch [13] illustrates this aptly:

> *It's amazing the debate is still going on ... Why is it, that in IT, we tend to hijack a word from another industry, and give it an altogether different meaning? ... The IEEE working group did a very good job in defining those terms back in 2000. Architecture is a property of a system reflecting its internal cohesion; its harmony with its surroundings and its design principles ... If only IT people would accept (and practice) the difference between an architecture description, a view as part of the architecture description (either made during the design of the system or afterwards) and the architecture itself, much confusion and mystification would be prevented.*

Following this line of reasoning, it can be argued that a distinction should be made between (a) the definition of architecture, (b) documentation of architecture, and (c) the use of architecture in practice (see Figure 1). The IEEE definition of architecture [14] states that architecture is a property of a system. In the case of EA, the system under consideration is an enterprise (thus assuming that architects have a systemic view of enterprises). The architecture of this system is its fundamental organization and the principles underlying this fundamental organization.

Paraphrasing the FRISCO report [15], it can be argued that since each architect has a unique perception of the enterprise under consideration, each architect sees the architecture of this enterprise differently (in FRISCO terms, an architecture is a special model which exists in the mind of an actor). Therefore, in order to gain a shared understanding of what constitutes this architecture, it is essential to communicate about architectures. This is also argued in e.g., the Archimate approach which introduces a modelling language to help facilitate easier communication about architecture between different (types of) stakeholders [10].

This brings us to the issue of architecture documentation which seems to be the focus of most of the big architecture frameworks and architecture languages such as IAF, Zachman, Archimate, et cetera. With respect to architecture documentation a distinction must be made between two schools of thought. First of all, there is the style of architecture blueprints in the form of (semi) formal diagrams which depict a possible (current or future) state of the firm in terms of processes, information flows, roles, information systems, infrastructure, et cetera. Based on recent experiences in discussions with managers it seems that this is what most people expect from architects: large plots of e.g. the application landscape.

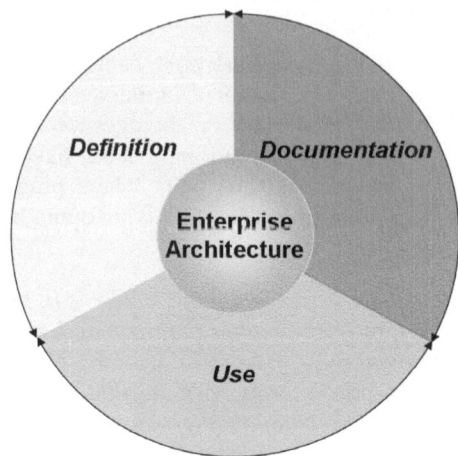

Fig. 1. Dimensions of enterprise Architecture

To a certain extent the Archimate language for architecture description is a proponent of this school of thought. In Archimate, the fundamental organization of (some aspect of) the firm is modelled in terms of static and behavioural elements (services). Frameworks such as IAF add to this the ability to reason about conceptual, logical, and physical services.

The second school of thought pertains to architecture principles (ignoring the distinction that is sometimes made between 'principles' and other types of regulations such as 'rules', 'guidelines', et cetera) which focus on restricting design freedom consistent with the xAF definition of enterprise architecture [9]. This school of thought seems to benefit a great deal from the business rules community [16,17].

Neither style is right or wrong. Both are equally valuable, depending on what one wants to achieve; the diagram style is predominantly descriptive in nature whereas the regulation style is predominantly prescriptive in nature. For example, the blueprints tend to provide a lot of information and insights that may be used for decision making whereas principles may be used in Project Start Architectures [18] to guide projects within the enterprise.

The last issue pertains to the use of enterprise architecture in practice. Based on several years of experience with enterprise architecture, [7] lists several key applications of enterprise architecture, most notably:

1. Situation description,
2. Strategic direction,
3. Gap analysis,
4. Tactical planning,
5. Operational planning,
6. Selection of partial solutions,
7. Solution architecture.

Putting these in a broader perspective, the book argues that the key point of enterprise architecture is to govern the process of enterprise transformation. Summarizing, it can be said that EA is about defining the desired future architecture of the enterprise, developing scenarios to achieve this future architecture, and governing the process of getting there; a process that is frequently dubbed *transformation* [19].

2.2 Strategic Management

Similar to the field of enterprise architecture, the field of strategic management is also characterized by many debates. These debates focus both on what strategic management is, and how strategic issues should be resolved in practice. In [20] a solid overview of the field of strategic management is given, that builds on the work of classical works on strategic management (such as [21,22,23,24]). For each of the well-known issues from the field of strategic management, the book presents two diametrically opposed perspectives and relates these to influential scientific material without passing judgement (.e., for the issue of competitive advantage, a resource-based perspective as well as a markets-driven perspective are presented). Due to the comprehensiveness and subjective nature of its analysis I will base my overview on the overview presented in [20]. The book makes a distinction between four dimensions of strategy: strategic processes, content, context, and organizational purpose (see Figure 2). Furthermore, different topics (pertaining to a strategic issue) can be distinguished in each of these dimensions. Summarizing, this boils down to:

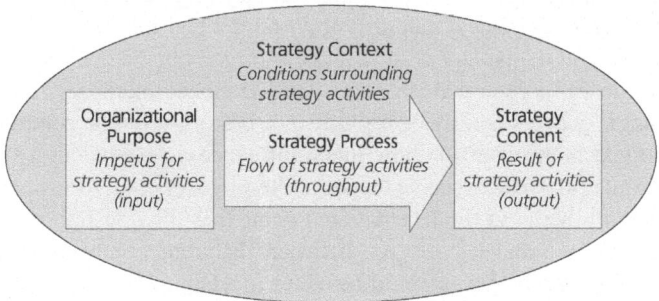

Fig. 2. Dimensions of strategy

1. **Strategy process:** pertains to the flow of strategy activities. This relates to the *how, who,* and *when* of strategy. The topics with respect to this dimension are strategic thinking, strategy formation, and strategic change. These three topics do not constitute entirely separate subjects; they are not phases, stages or elements of a process that can be understood in isolation.

   Strategic thinking deals with the question whether strategic processes are primarily rational (i.e., strategic alternatives are evaluated based on logical analysis, frequently involving cost-benefit analysis) or generative (i.e., in order to come up with the winning strategy it is necessary to think 'out of the box') in nature. Strategy formation deals with the question whether strategic processes are executed deliberately and lead to well-documented strategic plans, or whether strategy is considered to be emergent and can only be discovered in retrospect by analysing the decisions made by upper management. Lastly, the topic of strategic change pertains to the magnitude of change; should change be continuous (i.e., an evolutionary perspective) or discontinuous (i.e., a revolutionary perspective)?

2. **Strategy content:** pertains to the result of strategy activities. This relates to the *what* of strategy. The main question with respect to strategy content is positioning the organization with respect to its environment. Important questions for firms are: which arena do we want to compete in, and where in this arena do we want to be? In case of governmental organizations this implies that the organization must find ways to compare its performance to others (e.g., measure the performance of the Dutch police by comparing the number of murders solved with e.g. the police force of Belgium).

    Another aspect that is relevant with strategy pertains to synergies that the organization may strive for: at the level of the resource base, activity system, or product offering (see Figure 3). Finally, organizations should tackle the question whether they should develop their strategy in splendid isolation, or should move beyond a mere transactional relationship and develop strategy together.

3. **Strategy context:** pertains to the conditions surrounding strategic activities and relates to the *where* of strategy. The topics with respect to this dimension deal with growing contextual circles with respect to strategic initiatives. For firms these are the organizational context, the industry context, and the international context. For governments these levels can be transposed to a governmental agency (e.g., the police), a city (e.g., Rotterdam), or the entire country (e.g., the Netherlands).

    The organizational context can also aptly be dubbed the *leadership* context, and deals with the role of managers in achieving alignment with the environment and what input can be garnered from other organizational members? In other words, should strategic process mainly be top-down or bottom-up? The central theme in the industry context is the question how much influence a firm has in a specific industry. Is the firm a rule-maker (i.e., a leader in the industry) or a follower? Last but not least, the central theme in the international context is the question whether the firm should strife to maximize synergies between different countries with a standard product offering, or whether it would be better to adapt to local circumstances.

4. **Organizational purpose:** pertains to the impetus for strategy activities and relates to the *why* of strategy. In this case the main question is which concerns are leading with respect to the firm's strategy; should the firm seek to maximize shareholder value, or should the firm adopt a strategy with a high level of corporate social responsibility?

As can be seen from the previous discussion, different perspectives exist with respect to each of these topics / issues. These perspectives may even be diametrically opposed to each other. Each perspective is 'equally true', and provides valuable insights on how the underlying topics can be resolved in practice. In fact, managers are frequently presented with these diametrically opposed approaches to tackling an issue. Rather than arguing the case for one perspective over the other, the book takes a dialectic approach and suggests that the tension between perspectives should not be seen as an optimization problem, dilemma, or trade-off, but as a paradox where multiple innovative reconciliations of both perspectives should be considered. In the words of [25], the perspectives should be managed continuously as a polarity. We will return to the issue of polarity management in Section 4.

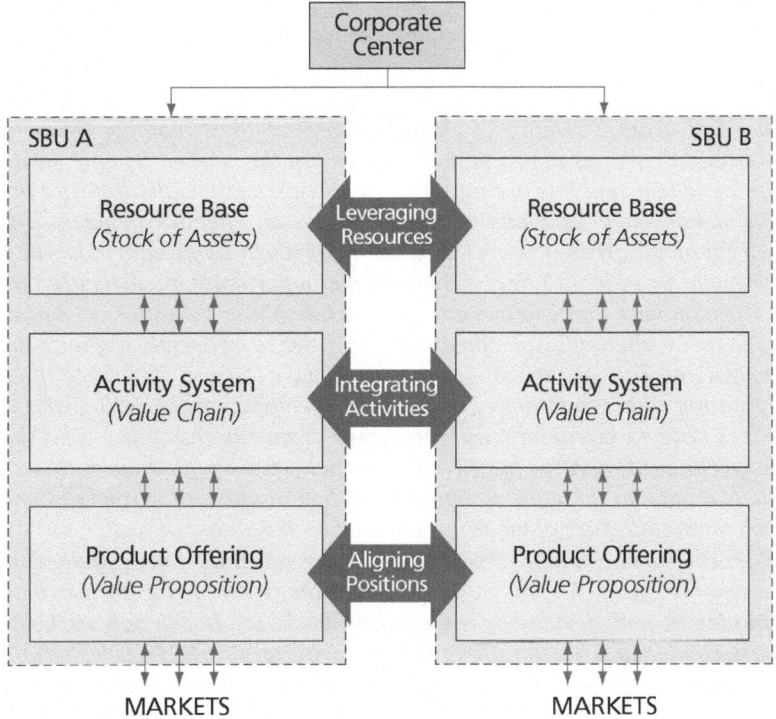

Fig. 3. Three types of synergies

3   Analysis

At first sight it may seem that strategic management and enterprise architecture attempt to steer the firm, yet focus on different aspects of the firm; strategic management attempts to control decision making related to the external position of the firm, in terms of alignment with the environment. Enterprise architecture attempts to exert forces on the firm with respect to its inner workings; its goal being to engineer the enterprise in such a way that it achieves its goal efficiently and effectively. If this were true, then both fields would form perfect *complements* and there would hardly be an issue at all. The following subsections list three main causes with respect to the tension between strategists and enterprise architects in practice.

3.1   Overlapping Domains

The first cause for a disturbed relationship between architects and strategists lies in the observation that the aspects with respect to the firm on which both fields have something to say overlap. This seems obvious, since the focus is on a single firm. Indeed, in many organizations, strategic decisions (frequently made by upper management, supported by staff from a strategy department) impact the fundamental organization of (the working of) the firm. In other words, strategic decisions may require transforming the firm into a new 'state' which is best achieved by governing the process under architecture (see Section 2.1). The following examples illustrate this:

Example 2

- *Assume that a large retailer of electronic devices (most notably computer hardware) traditionally has aimed at the lower end of the market. Its core strategy is to be a price leader and has organized its processes ...especially logistics and information management surrounding logistics ...in an effective manner. Due to the current financial crisis it perceives a unique growth opportunity: it will target a new segment in which it offers customers the opportunity to customize their products. The company wants to out-source the assembly to a partner organization and leverage its excellent logistic capabilities in order to efficiently organize this mass-customization. This change in strategy may have serious impact on the desired configuration of the enterprise. In order to effectively organize logistics, internal processes have to be aligned with processes of an external party. The same goes for supporting information systems, posing new issues with respect to security, et cetera. Not only do these aspects have to be investigated, the principles underlying the current architecture of the firm should be reconsidered as well.*
- *Consider the same company 5 years down the road. It has successfully entered the new market segment. As part of its growth strategy it wants to further build on its capabilities of and decides to enter a new business. It will sell its home-grown software combined with consulting services with respect to logistics to other companies. Again, prospective changes for the fundamental organization of the firm are potentially huge. Next to entering a new business which requires new staff, new processes, information systems, policies, et cetera, IT staff is suddenly confronted with possibly conflicting requirements from prospective new customers. Even more, working according to a strict release schedule might become increasingly important.* □

As can be seen from these examples, a change in strategic direction of a firm may lead to a transformation of the firm, and thus to its architecture. Governing this transformation is, as we've seen, part of working under architecture. One of the core assumptions that enterprise architects tend to make in practice, is the fact that the strategy of the firm is a known, and is stable enough to craft an architecture for. However, changes to strategy appear to be more frequent than sometimes hoped. For example, [26] lists several strategic changes to the resource base of a firm (ignoring reasons to make strategic changes to the three other key aspects of a business model: the activity system, product offering, and revenue model):

- The opportunity to address through disruptive innovation the needs of large groups of potential customers
- The opportunity to capitalize on brand-new technology by wrapping it in a novel business model
- The opportunity to bring a job-to-be done focus; i.e., customer-driven rather than product centric
- The need to fend off low-end disruptors
- The need to respond to a shifting basis of competition

Example 2 illustrates the fact that changes in strategy can be both sudden and unforeseen (who suspected in late 2007 that the world would suffer from a financial crisis in 2008?) and of considerable magnitude. Frequent and unforeseen changes of this kind are partly the reason why strategists sometimes have a bad name with architects. Similarly, failure to make real progress in re-engineering the firm (especially in terms of supporting IT) frequently gives architects the name of being inflexible and ignorant of business issues.

So far the discussion has focussed on the impact of strategic change to the (desired) architecture of the firm. However, the reverse is also possible: having a solid enterprise architecture may influence strategy formation. To be more specific, enterprise architectures could potentially play an invaluable role in strategy formation processes. To see why this is true, consider the following example:-

Example 3 - *Consider a multinational company that sells magazines (print), but also is active on the Web. Magazines tend to have websites of their own (country-based), but other websites of said company are targeted at an international audience. Aiming at reaping synergy benefits for its web business, the firm explores its strategic options. These options are evaluated in terms of market attractiveness and resource-base fit. The role of architects in this process could be:*

- *Give insight in feasibility of certain options in terms of resource-base fit*
- *Help generate options based on extensive knowledge of current operations (inside-out perspective)*
- *Provide realistic scenarios in terms of planning, making analysis of when (timing) initiatives start to pay off more realistic.* □

In summary, strategists and architects could in theory benefit greatly from each other. In practice, however, it seems that it is insufficiently understood that strategy and architecture are complements, dealing with similar aspects of the firm. As Tom Graves (Principal Enterprise Architect at Tetradian Consulting) puts it:

> *It is impossible to do real enterprise-architecture where the "enterprise architecture" unit is dominated by IT, or is under IT governance alone. Functional enterprise-architecture only becomes possible when it has a true enterprise-wide scope, reporting to the CEO or some other enterprise-wide body that has whole-of-enterprise authority. At that point it not so much has a relation with strategy, as that it is strategy and a means of expression of strategic outcomes.*

Rather than argue over 'who is in control' the two disciplines should work together to strengthen the firm as will be argued shortly.

3.2 Different Language

It seems that strategists and architects tend to use different language; using different labels to address some concept, as well as re-interpret concepts differently. Perhaps the most prominent example of the latter is the word *strategy* itself. In several organizations I have observed that the word 'strategy' is equated to 'plan'. For example, at vtsPN[4], the

---

[4] http://www.vtspn.nl

organization which helps the (semi)autonomous Dutch police forces as well as other parties involved with public safety and healthcare, the strategy consisted of a voluminous plan to restructure and optimize the application portfolio of the 26 Dutch police forces. Similar issues exist with concepts such as 'model', 'framework', 'system', 'innovation' et cetera (Interestingly, within the field of IT a similar situation existed in the late 1980's and early 1990's. This resulted in several attempts to create a clear definition of commonly used terminology within this field. The FRISCO framework is a prominent example of this [15]). The following anecdote illustrates how misinterpretation between strategists and architects contributes to poor relations between strategists and architects.

Example 4 – *Consider once more the retailer from Example 2. After successfully transforming the company due to starting a new business (selling software combined with consulting), the company holds its annual strategic planning session. Overrun of the transformation process in terms of budget and planning was minimal and the process is considered to be a big success. For the first time in years the company's chief enterprise architect is invited to join the session. The CIO argues the case for reaping more synergies at the level of the resource in order to make the company as a whole more agile.*

The enterprise architect, having successfully led the transformation, reacts by saying that the new processes, information flows, supporting information systems and infrastructure have only just been put in place, humming like a well-oiled machine. Between the lines, he suggests that further optimization hardly seems possible. Several members of the strategy team misinterpret this remark, which seems to suggest that the board member's ideas are unrealistic. Before the situation gets out of hand, the CIO explains that the idea is not to optimize processes or reconsider the application portfolio. Instead, the goal is to reuse knowledge and capabilities across businesses.   □

This example shows that misunderstanding may lead to tension between both groups of practitioners. Misunderstanding is not so much the issue; the consequences from misinterpretation can be, though.

3.3 Different Worldviews

It seems that, in general, strategists and architects tend to have different worldviews with respect to organizations, similar to the observations made in [27]. Strategic management is a socio-economic topic and as a result the predominant view of strategists with respect to the firm is organismic / societal. On the other hand, enterprise architecture is generally considered to be an engineering discipline in which the predominant view with respect to the firm is systemic. This can readily be seen from the IEEE definition of architecture [14]. In this light the work of Stacey is particularly interesting, as [28] provides a systemic view of the firm and relates this to the field of strategic management.

In the organismic worldview, the firm is seen as a society of individuals; these individuals organize themselves and *are* the firm. These individuals opt to join forces in (changing) alliances to achieve common goals. As such 'the firm' can be seen as a social construct, a vehicle that assists in achieving goals of its participants. The strong point of this view of the firm is that it successfully explains the non-deterministic behavior of the firm, taking into account the political aspects related to any type of society. Even more, the work of Simon has greatly improved our knowledge of how decision making in societies take place, particularly in uncertain situations (e.g., [29]). In this view, however, the boundaries of 'the firm' seem unclear and, due to the focus on social constructs, issues such as efficiency of the firm in achieving the goals of all its participants are difficult to measure / analyze.

The perspective of the systems' worldview has as its strong point that the firm is seen as a deterministic system; free to shape as one desires. Manipulation of the enterprise is seen as complex, yet relatively predictable. Human behavior / preferences are mostly ignored in this perspective which leads to reduced complexity. This has the advantage of clear lines of control, easier governance and a clear boundary of what constitutes the organization. However, due to the fact that human behavior is mostly ignored, the politics that are present in organizations are largely ignored (e.g., shirking). Even more, inertia is also ignored which may makes decision making harder than is actually assumed, e.g., because it is unclear when certain measures will start to actually affect people within the organization.

It seems that these differing worldviews mainly exist due to the way our educational systems are organized; strategic management is typically taught at economics departments and business schools whereas enterprise architecture is mainly taught at informatics and computer science departments. It remains to be seen whether it makes more sense to teach enterprise architecture as a joint venture between economics departments and computer science / informatics departments to alleviate some of the pressure. It may very well be that the other two causes for the tension between strategy and enterprise architecture that have been identified in this paper are a direct result of the fact that the worldviews of participants differ. The following example illustrates how the two differing worldviews lead to different conclusions in a situation in practice.

Example 5 – Consider the situation in which a firm has a portfolio of 25 semi-autonomous business units in separate geographical locations. Each unit has its own processes and information systems which are very much alike. In some cases these systems once were duplicates, but have evolved differently over the last few years. In this situation, it makes sense from an engineering (architecture) point of view to strive for standardization, especially in terms of information systems. This will ease the maintenance burden, makes it easier to implement new requirements from top management (lower development cost), and even makes it easier for personnel to switch from one region to the next. Considering this situation from a societal perspective, however, may lead to the insight that the local businesses may be attached to their systems and may not want to switch systems due to organizational inertia, political reasons, or even loss of specific (local) functionality.   □

4 Resolving the Tension between Strategists and Enterprise Architects

Given the tension between the worldviews 'the enterprise is an organism' and 'the enterprise is a system', the question for managers is how to deal with these different perspectives. In polarity management [25] it is asserted that managers should continually manage this issue and strive to use the best points of both worldviews. In [20] a more elaborate view on this question is presented. In general there are three ways to deal with tensions: issues are seen as a dilemma, a trade-off or a paradox. The following list summarizes these views on tensions and illustrates them with (IT-related) examples:

- As a *dilemma*: a dilemma is a vexing problem with two possible solutions. In dealing with a dilemma, both 'sides' will enter a debate an attempt to convince the other of their wrong. In practice it turns out that the pitfalls of the chosen perspective will manifest, leaving the decision maker with the uneasy feeling that the upside of the perspective that is not chosen are ignored, despite the good arguments in favor of this side. It may be tempting to 'switch sides' and conflict between two camps is often the result.

  Consider the situation where the entire IT-portfolio of a firm is to be redesigned. With several business units at different geographical locations, both a centralized and decentralized approach are considered. When this issue is considered to be a dilemma then decision makers must *choose* one alternative over the other. Both alternatives may have advantages and disadvantages which must be weighted in the decision. Frequently, however, the decision cannot be made on logical grounds alone. The supporters of both sides of the dilemma will argue their case and strive to gain the upper hand. Satisfying one group may lead to disappointment of the other group.

- As a *trade-off*: a trade-off is a problem situation in which there are many possible solutions, each striking a different balance between two conflicting pressures (in the Netherlands this is often called 'polderen'). In this mode of decision making both sides exchange pro's and con's of their perspective and the decision maker is left with the task of selecting a solution that is acceptable to both sides. Often this leads to sub optimization in the form of a compromise.

  Continuing the previous example, when the issues is perceived to be a trade-off then not only the extremes (a centralized systems versus a decentralized system) are considered, but also trade-offs between the two. This may, for example, lead to a situation where each business unit has a *locally centralized system* that is the same for all business units. These local systems are interconnected with a service bus for easier integration.

- As a *paradox*: a paradox is a situation in which two seeming contradictory, or even mutually exclusive, factors appear to be true at the same time. A paradoxical problem has no real solution, as there is no way to logically integrate the two perspectives on the paradox. If this approach is taken, the conflicting between the two opposites is accepted; the two 'sides' enter a dialogue in which they attempt to solve the issue. The decision maker is left with the task to accommodate (the upsides of) both perspectives at the same time (e.g. [25]). A resolution to a paradoxical issue is dubbed the synthesis of this issue.

  Continuing the previous examples, then the issue is considered to be paradoxical a synthesis must be sought. In this case, one seeks for a situation that has the upsides of both sides of the polarity. This may lead to a situation where the centralization/decentralization polarity is constantly managed by systems-reconfiguration depending on the needs from a specific business unit at a specific time.

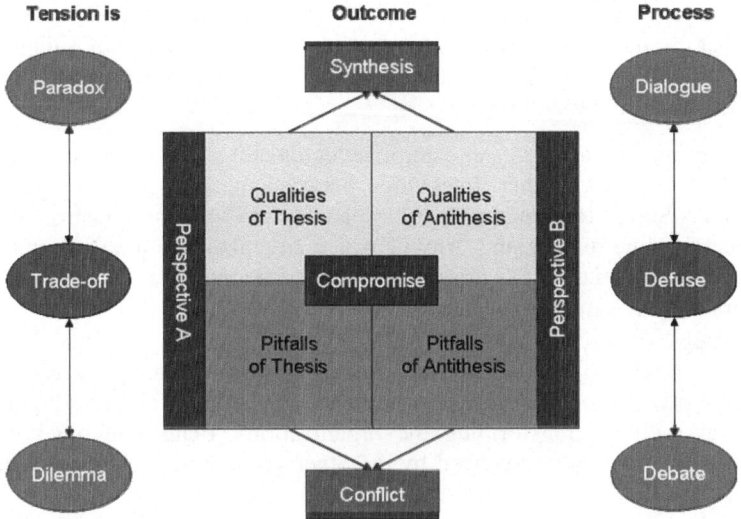

Fig. 4. Three types of problems + resolutions [20]

The different modes of dealing with tensions are summarized in Figure 4. As can be seen in this figure, when the issue at hand is considered to be a dilemma where one has to choose between alternatives then the outcome is likely to be a conflict where the pitfalls of both options manifest. Similarly, when the issue is seen as a trade-off then the end result is likely to be a compromise; a sub-optimization which combines some of the qualities and pitfalls of both sides of the issue. When the issue is seen as a paradox then the qualities of both sides can combined to form a synthesis, avoiding the pitfalls of both sides.

Firms should strive to find a synthesis between the two worldviews discussed in this paper to find an acceptable resolution for the tension between business and IT. That is, they should find a synthesis between the worldviews 'the enterprise is an organism' and 'the enterprise is a system'.

The phrase "finding a synthesis" suggests that the issue of reconciling strategy and architecture is a one-shot issue; something that can be resolved by making the right decision at a certain point in time. This is in fact not the case. The issue of relieving the tension between strategists and enterprise architects is a *polarity* that must be managed. According to [25], a polarity is characterized by two things:

1. *The difficulty is ongoing*: implying that no single decision or change can solve the issue. The issue needs constant 'solving', especially since the issue does not seem to have a definite endpoint (solution).
2. *The two poles are interdependent*: implying that the two 'sides' can not stand alone, as is the case in e.g., a dilemma. In case of the issue at hand, it may seem that the two perspectives *can* stand alone. However, as has been argued in this paper (see Section 3.1), the two perspectives complement each other.

Adhering to this view implies that the phrase "the tension between the fields of strategic management and enterprise architecture must be solved" is more accurately phrased "the tension between the fields of strategic management and enterprise architecture must be managed". Firms should therefore look for *social* solutions in the sense that the two communities should be brought together. A practical 'solution' for a firm might entail the following aspects:

- The two groups (strategists and enterprise architects) share an office, or are at least situated close to each other. Interaction between the two groups is paramount to success. As Sumar Ramanathan (senior manager at Capgemini, Detroit area) stated in a recent online discussion: "Any IT or EA organization that is disengaged with business is doomed to fail as proven by many catastrophic failures expressed as big bloopers of IT/EA management in global enterprises"
- Both groups should work together as frequently as possible. That is, enterprise architects should be involved in the strategy formation processes, and strategists should be involved in the process of engineering the firm (i.e., developing end solutions as well as transforming the organization). Plans and other documents should be formally peer reviewed by the other group before they are submitted to e.g., the board.
- To ease the burden of communication, the two groups should learn the lingo of the other group. This can be achieved by taking (joint) courses, organizing frequent colloquia, sharing literature, et cetera. Note that strategists should not have to *become* architects (or vice versa), the goal is to improve communication, understand the issues that are top of mind, et cetera.
- Both groups should *jointly* be rewarded for successes, as well as commented on when failing.

These suggestions are in line with common business practices. For example, [30] accounts of the successful architecture initiatives at ABN AMRO over the last ten years. Key success factors that are reported are trust, actively participate in strategic sessions, orchestrate frequent sessions for discussion the architecture and keeping it lean and mean.

5 Conclusions and Future Research

It seems that the tension between 'business' and 'IT', which has been around ever since computing entered the business realm, has been transposed to a tension between strategists and enterprise architects. This seems especially true in firms where enterprise architecture is directly linked / associated to pure IT.

Analysis of the theory of both fields suggests that the fields of strategic management and strategy are complements; strategic management focuses (mostly) on the relation of the firm to its environment, thus setting a course for the firm, whereas enterprise architecture focuses on developing scenarios that conform to the desired strategy as well as governing the transformation process of the firm to actually achieve its strategic goals. If this were true, and all humans would be completely rational, then the tension should not exist.

Reality is different, however. In this paper I have identified three ground causes for the tension between the two groups of practitioners. The first issue has to do with the fact that the two fields are more interrelated as might seem at first sight. Changes in strategic direction (which can be the result of a variety of issues) tend to result in the need for change in an organization which is one of the strong points of the field of enterprise architecture. Similarly, having a solid enterprise architecture may effectively be (part of) a firm's competitive advantage. It seems that in many cases, strategists and architects perceive this overlap in domains to be a dilemma of 'who is in control', which does not improve the working relation between the two groups of practitioners.

The second ground cause lies in the fact that the two groups speak a different language: using different 'labels' to address the same concept. This is partly the result of the two groups having different concerns (i.e., different issues with respect to the firm are top of mind), but also due to the fact that the two communities hardly overlap: they share little formal education, hardly read literature with respect to each other's field, et cetera.

The third ground cause, which seems to preclude the two other causes for the tension, lies in the fact that the predominant worldview of both groups differs. The predominant worldview of strategists seems to be organismic / societal in nature, whereas enterprise architects have an engineering perspective with respect to the firm. These views with respect to the firm are diametrically opposed, which also points in the right direction with respect to resolving the tension between strategic management and enterprise architecture in practice.

I have argued that the tension between strategy and architecture should not be perceived as a dilemma or trade-off. Instead, the tension should be perceived as a paradox, and there is no single decision that can resolve it. Instead, the tension should be managed, conform the idea of *polarity management* as proposed by Johnson [25]. To relieve the tension in a single firm, the two groups should seek to join forces in an attempt to jointly strengthen the firm. Strategists should not have to become architects (and vice versa), but a shared understanding, and resolve to face, the challenges of the firm should be strived for. In the long run, i.e., across firms, the tension between the two fields can only be resolved when the field of enterprise architecture matures and gains respect by repeatedly adding value for firms.

As was stated in the introduction, hard empiric evidence on many topics related to the relation between enterprise architecture and strategic management are missing still. It seems worthwhile to conduct a survey among organizations (both public and private, in the Netherlands as well as abroad).

References

1. Booch, G., Maksimchuck, R., Engle, M., Young, B., Connallen, J., Houston, K.B.: Object-Oriented Analysis and Design with applications. Addison-Wesley, Reading (2007)
2. Bijl, D.: Service Orientatie en ICT. Microsoft Press (2005)
3. van Es, R., van Gerwen, N., Graave, J., Lighthart, A., van Rooij, R.: Flexibel omgaan met voortdurende veranderingen - een praktische leidraad voor het invoeren van een Service Oriented Architecture, Ordina (2005)
4. Parker, M., Benson, R.: Enterprisewide Information Management: State-of-the-art StrategicPlanning. Journal of Information Systems Management, 14–23 (1989)
5. Henderson, C., Venkatraman, N.: Strategic alignment: Leveraging information technology for transforming organizations. IBM Systems Journal 32(1), 472–484 (1993)

6. Zachman, J.: A framework for information systems architecture. IBM Systems Journal 26 (1987)
7. Op't Land, M., Proper, H., Waage, M., Cloo, J., Steghuis, C.: Enterprise Architecture - Creating Value by Informed Governance. Springer, Heidelberg (2008)
8. Ross, J., Will, P., Robertson, D.: Enterprise Architecture as Strategy - Creating a foundation for business execution. Harvard Business School Press (2006)
9. xAF working group Extensible Architecture Framework version 1.1 (formal edition) NAF (2006)
10. Lankhorst, M. (ed.): Enterprise Architecture at Work: Modelling, Communication and Analysis. Springer, Heidelberg (2005)
11. The Open Group: The Open Group Architecture Framework (TOGAF). Van Haren Publishing (2007)
12. Hendriks, O.J. (ed.): Architectuur maakt de toekomst mogelijk - Landelijk Architectuur Congres 2008. Academic Service (2008)
13. Bosch, J.: Comment on three perspectives on Enterprise Architecture. Weblog of Erik Proper (2007)
14. IEEE Computer Society: IEEE Std 1471-2000: IEEE Recommended Practice for Architecture description of Software-Intensive Systems (2000)
15. Falkenberg, E., Hesse, W., Lindgreen, P., Nilsson, B., Oei, J., Rolland, C., Stamper, R., Assche, F.V., Verrijn-Stuart, A., Voss, K.: A Framework of Information Systems Concepts. IFIP WG 8.1 Task Group (1998)
16. Business Rules Group: Business Rules Manifesto Business Rules Manifesto (2003)
17. Object Management Group: Semantics of Business Vocabulary and Business Rules (2006)
18. Wagter, R., Berg, M.v.d., Luijpers.: J. DYA: snelheid en samenhang in business en ICT architectuur, Tutein Nolthenius (2001)
19. Gouillart, F., Kelly, J.: Transforming the organization. McGraw-Hill, New York (1995)
20. deWit, B., Meyer, R.: Strategy Synthesis, Revolving Strategy Paradoxes to Create Competitive Advantage - Concise version. Thomson (2006)
21. Porter, M.E.: Competitive Strategy: Techniques for Analyzing Industries and Competitors. Free Press (1980)
22. Porter, M.E.: Competitive Advantage: Creating and Sustaining Superior Performance. Free Press (1985)
23. Mintzberg, H., Ahlstrand, B., Lampel, J.: Strategy Safari: A Guided Tour Through the Wilds of Strategic Management. Free Press (1998)
24. Kaplan, R.S., Norton, D.P.: The Strategy-Focused Organization: How Balanced Scorecard Thrive in the New Business Environment. Harvard Business School Press (2001)
25. Johnson, B.: Polarity management - identifying and managing unsolvable problems. HRD Press (1996)
26. Johnson, M.W., Christensen, C.M., Kagermann, H.: Reinventing your business model. Harvard Business Review 86(12), 51–59 (2008)
27. Snow, C.P.: The Two Cultures. Cambridge University Press, Cambridge (1993)
28. Stacey, R.: Strategic Management and Organisational Dynamics, 5th edn. Prentice-Hall, Englewood Cliffs (2007)
29. March, J.G., Simon, H.: Organizations, 2nd edn. Wiley-Blackwell (1993)
30. Schmitz, W.: Tien jaar architectuur bij ABN AMRO. Informatie 50(9) (2008)

Measuring the Risks of Outsourcing:
Experiences from Industry

Michiel Borgers[1], Frank Harmsen[2], and Linda Langheld[3]

[1] Former Capgemini, Sourcing Consulting, Utrecht
m.a.c.borgers@minfin.nl
[2] Maastricht University, Maastricht and Capgemini Sourcing Consulting, Heerlen
frank.harmsen@capgemini.com
[3] Capgemini Sourcing Consulting, Utrecht
linda.langheld@capgemini.com

Abstract. In this paper we present the findings of 50 cases regarding the application of the Rightshore Assessment Study (RAS), a tool and approach for determining the sourcing scenario based on the measurement of the risks of outsourcing. RAS is aimed at assessing both the organizational capability and the complexity of the IT domain, yielding the potential qualitative and quantitative benefits and risks for outsourcing information systems portfolios or part thereof. After a description of RAS, we present an overview of the findings of the studies of the past few years. These studies have been conducted at 10 different organizations in various industry sectors. We have related aspects such as process maturity, people, knowledge management, IT platforms and support to the benefits and risks of outsourcing. In this paper, we compare these findings, analyze them and draw conclusions on both the outsourcing risk factors and ways how to improve them. We conclude with a chapter with an analysis of the approach and ideas for further research.

Keywords: Sourcing strategies, risk management, enterprise transformation.

1   Introduction

Recent studies (e.g., [3], [4]) show, in most markets and countries IT outsourcing and its many different variants (which we summarize under the heading 'sourcing') is still on the rise. Organizations that have outsourced parts of their IT, have experienced a higher degree of IT agility. In the past few years, a number of trends have been signaled:

- IT governance is gaining importance, in particular in relation to the so-called *retained organization* that manages business demand and translates this demand to the governance of the sourcing partners. In recent years, both in the afore mentioned studies and in scientific literature [8], much attention has been dedicated to the capabilities such a retained organization should possess. IT governance enables *multisourcing* [2], which requires a higher degree of maturity of the retained organization [1].

E. Proper, F. Harmsen, and J.L.G. Dietz (Eds.): PRET 2009, LNBIP 28, pp. 197–209, 2009.
© Springer-Verlag Berlin Heidelberg 2009

- Sourcing is globalizing. Because of the tight labor market, the talent pool in emerging countries, the cost/benefit ratio and the technological developments enabling globally distributed and location-independent work, larger organizations in particular tend to distribute their IT function around the globe.
- IT budgets are still under pressure, and this trend will, given the current economical crisis, only be stronger the coming years. Not only is the business requiring lower IT costs, these costs should also be more predictable. For this reason, organizations seek suppliers that are willing to invest heavily upfront, in order to gain quick wins and reduce IT costs and increase predictability on short term. So-called *Master Vendor* and co-sourcing contracts are on the rise to substantiate such sourcing partnerships.

Given the fact that there are many variants with respect to geography, business model, distribution model, responsibility sharing, etc., organizations have many options regarding their IT sourcing strategy and the way this strategy is implemented through various scenarios [6]. Every organization has its own characteristics and in order to achieve maximal effectiveness, these characteristics need to be taken into account as much as possible [7]. Already in a small region like the Benelux, for instance, we see in our client base a large variety of factors that determine what and how to source. This is dependent on the sector of an organization (for instance, the electronics industry is outsourcing virtually anything to emerging countries), but also on region (in the southern part of the Benelux organizations tend to be more careful with outsourcing due to political sensitivity, and they are choosing co-sourcing constructions) and on the maturity and international focus of an organization (a multinational corporation is used to sourcing work to other countries, but for a local health insurance company this is less obvious). Often, these factors are taken into account to determine a sourcing strategy and various sourcing scenarios [6], but it is hard to trace back sourcing decisions to the specific situation factors of an organization. This increases the chance of sub-optimal or even bad decisions, mostly because of pre-occupation of organizations towards these decisions.

The decision problem gets even more complicated if the various functions of the organization are taken into account. Different functions deserve different sourcing models. And even *within* an organizational function, different strategic choices and scenarios with respect to sourcing are needed to achieve an optimal sourcing situation. In our paper we focus on the differentiation of possible sourcing alternatives for an organizational function. The instrument we have been using in a number of real-life cases is oriented on the IT function, but the philosophy that underpins it can be used for other functions as well.

In order to enable organizations to choose, in a motivated way, from different scenarios, we have developed the Rightshore Assessment Study (RAS). We have been using this outsourcing readiness study to help clients with their sourcing scenario decision making, in particular with respect to their IT application portfolio. The philosophy of this readiness study is that outsourcing risks are based on the complexity of an IT application and the capabilities of the (retained) organization. Both Application Complexity and Organization Capability are measured using a set of variables, also called outsourcing risk factors. The assumption is, that outsourcing is a relatively high risk activity – if lower risk is needed, other sourcing models have to be considered. RAS scores characteristics and assists consultants in providing sourcing scenario suggestions.

In a former paper we addressed the impact of operational risk occurrence in offshore application development projects and which mitigation actions were most effective [5]. That research was focusing on risks in offshore application development described in literature and validated by experts in the work field, like project managers and sourcing consultants. One of the most important results was the fact that risks from organizational origin have higher impact than risks from an application origin.

In this paper and a following paper we present findings of 50 cases of our Rightshore Assessment Study. We use the available data from practice to try to understand the relationship between outsourcing risk factors and advised sourcing scenarios. Out of the findings we want to answer the following question:

'What is the impact of the different outsourcing risk factors
on the sourcing scenario advised to the client?'

The hypothesis behind this question is the idea that the higher the risks the less amount of work is outsourced to a supplier.

In the following section we start with an overview of the RAS tool. After that we analyze the risk scores of the 50 cases and discuss the approach used. We end this paper with conclusions and suggestions for further research.

2 RAS Overview

The Rightshore Assessment Study is a methodology used by an IT consultancy firm. The objective of RAS is to define possible sourcing scenarios for an organization, including the related financial business cases. A sourcing scenario is determined based on the insights in the outsourcing risks, sourcing strategy preconditions and the way in which the Application Development (AD) and Application Maintenance (AM) activities are organized.

For each sourcing scenario a financial business case can be calculated to define the possible cost savings over a period of time. Current costs are calculated, costs for the transition and ongoing costs for the future scenario. Both transition costs and ongoing costs are specific for a certain sourcing scenario. Important for calculating the business case is to know what amount of investments are necessary to mitigate specific outsourcing risks, since they influence the transition costs and possibly the future ongoing costs.

So for both determining the sourcing scenario as for calculating the business case the measurement of the outsourcing risks are key. The outsourcing risks are based on two pillars: Application Complexity and Organization Capability. For each pillar several outsourcing risk factors are defined based on specific characteristics of the IT application and the related organization.

2.1 Application Complexity

Pillar one of the outsourcing risks analysis is the Application Complexity. The IT application is the subject of the outsourcing and therefore the hypothesis is that characteristics of the application will influence the final sourcing scenario. To quantify the effect of the application on the outsourcing risks, seven Application Complexity risk factors are defined (see table 1 for the descriptions).

Table 1. Application Complexity Risk Factors

Risk Factor	Description
Architecture	Availability and use of (data, application, technical) architecture and the characteristics of the applications.
Business Logic	Size and different types of business functions supported by the application and the criticalness of the application.
Application	Generic characteristics of the application, like size, type of technologies, amount of change requests, etc.
Interface	Size and types of interfaces related with the application.
Infrastructure	Hardware and software necessary to manage and to operate the application, like servers, network, operating systems, DBMS, etc.
Interaction	Level of interaction during the different activities of AD and AM between engineers and with end users.
Support	Level of support needed for this application, like support requirements as language support, 7 x 24 hours, type of communication with end users.

Per risk factor specific characteristics are investigated. During the RAS it is determined to what extent the characteristics will contribute to a higher outsourcing risk or not. For example ´the number of annual change requests per year´ is determined as one characteristic of the risk factor ´Application´. A high number of annual change requests, will increase the risk factor. All those characteristics are weighted with the same value to the risk factor and all factors are weighted with almost the same value to the overall Application Complexity score[1].

2.2 Organization Capabilities

Organization Capabilities is the second pillar of the outsourcing risk analysis. The idea behind this pillar is the fact that organizations should have specific capabilities to govern the outsource supplier. If the capabilities of the organization are all in place, the risks during outsourcing are lower than when some capabilities are missing. We have defined five risk factors to determine the capability of the specific organization (see table 2).

2.3 Risk Overview Matrix

When both the Application Complexity and Organization Capability are determined, the scores can be plotted in the Risk overview matrix (see Fig 1.). When the Application Complexity is low and the Organization Capability is high, the outsourcing risks are

---

[1] All factors are weighted with 15% and only 'Support' risk factor is weighted with 10%.

relatively low. The organization is capable to govern the supplier and processes like knowledge management and requirements management are in place, which lower the risk of the transition. A low application complexity also positively influences the complexity of the transition and therefore the risks of outsourcing. The opposite might occur as well, where the complexity of the application is high and the capability of the organization is low. In this case, there might still be reasons to outsource the application, but there will be more challenges to overcome. Before or during transition to the external supplier those risks can be mitigated by rebuilding the application to lower the complexity or by taking organizational measures to increase capabilities. Of course the application can be outsourced without any mitigations actions as well, but then the advise will be to save some risk money.

Table 2. Organization Complexity Risk Factors

Risk Factor	Description
Knowledge Management	Availability of Knowledge Management processes and tools and the level of completeness of all kinds of documents (like requirement specifications, design documents, etc.).
Methodologies & Tools	Level of use of methodologies and tools to be used during AM/AD activities.
Testing	Use of the different types of testing during AD / AM activities and the way testing is managed.
Requirement Management	Level of specification of requirements and the availability of a Requirement Mgt. processes to manage the requirements.
People	Level of knowledge and training available in the retained organization.

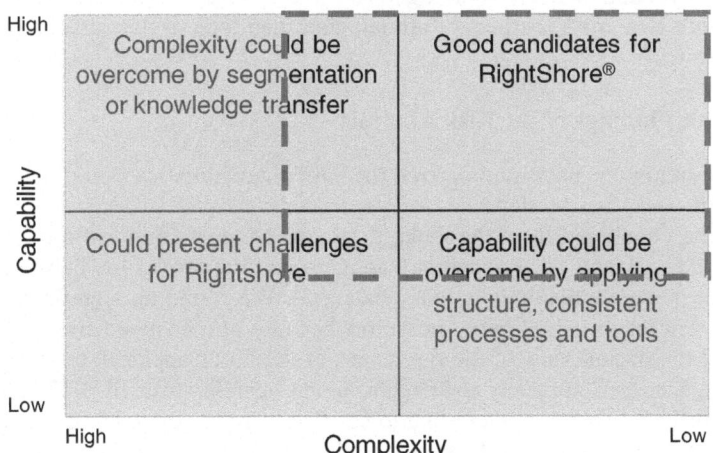

Fig. 1. Risk overview matrix

2.4 Outsourcing Risk Measurement Approach

To determine the scores of all the risk factors for both Application Complexity and Organization Capability a scorings questionnaire has been used with predefined answers and a scoring value. For Application Complexity we have a total of 97 questions with an average of 13 questions per risk factor and for Organization Capability we have a total of 61 questions so an average of 12 questions per risk factor. For each question the possible answers are ranked between 1 and 5. Some of the questions are objectively defined, e.g. how many external interfaces has this application, and other questions are subjective: how complex is the application architecture: high, medium or low.

To be able to fill in the scorings questionnaire, data is collected in three different ways: interviewing, document study and own research. All data collected is compared and analyzed by a consultant who determines what the final answers to the questions are. After answering all the questions the average score per risk factor, and also the overall average score for the Application Complexity and Organization Capability are calculated.

3 Analyzing the Risk Scores

We have applied RAS to 50 IT applications, belonging to 10 different organizations. Using RAS, we identified for each IT application the Application Complexity and the Organization Capability. Each case has its own characteristics with respect to:

- The industry sector: Banking & Insurance, Electronics, Pharmaceutical, Telecom, or Transport.
- Application type: Package-based or custom made.
- The IT function: Application development or application management.

In this section we will analyze the data in different ways to come up with interesting findings related with outsourcing risks. We first start with an overall analysis, continue with two cross sections of all data and then look at the outsourcing risk factors per pillar.

3.1 Overall Findings of the Risk Analysis

In the table below we have summarized the case characteristics using the sector of the organization as a cross section.

As can be concluded from the table, the focus of most cases is on the Banking & Insurance industry and on application management. The proportion between packages and custom made applications is fairly balanced. We found no significant difference in average risk factor values between sectors because of the spread size of the data.

In Fig. 2 the overall view of the risk scores of the 50 IT applications is depicted. The average Application Complexity and Organization Capability over all 50 IT applications is 3.0 respectively 3.1, but it is interesting to see that there are three IT applications with a relatively low overall risk score. Reason for those low scores is the size of the applications: they score low with respect to many characteristics on both technological and business related elements. It is understandable that when an application is used for a small amount of users and a small amount of business functions the technological implementation of that application will be small as well and therefore the complexity is low. And when an application has a low complexity than it should be easier to support that application from an organizational point of view.

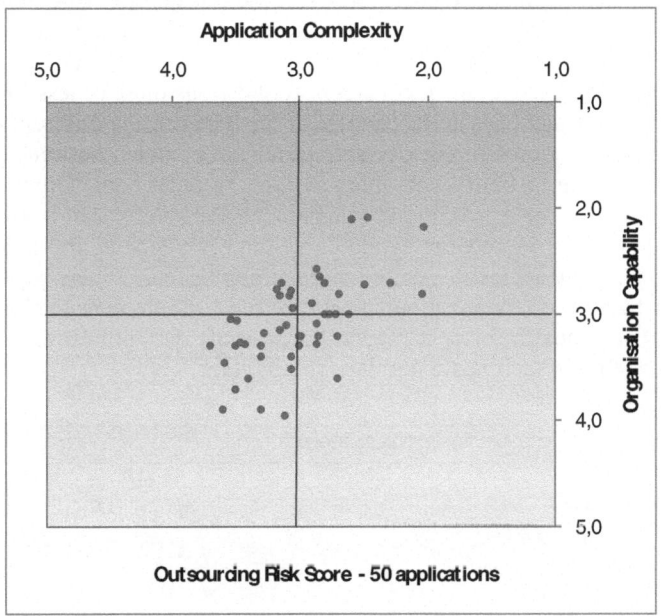

Fig. 2. Risk overview matrix

Table 3. Overview of case characteristics of the 50 IT applications

Sector	Package	Custom	AD	AM	Average complexity	Average capability
Banking & Insurance	12	22	6	28	3.0	3.0
Electronics	3	0	0	3	2.8	2.9
Pharma	1	0	1	0	2.9	2.6
Telecom	4	4	0	8	3.1	3.2
Transport	1	2	0	3	3.3	3.1

3.2 Cross Sections

We have looked at two cross sections in particular, because we would expect that there might be a difference in the level of risks for outsourcing as we will explain later. One is the difference of IT function to be outsourced, the other is the type of application.

Application Management vs. Application Development
We were interested in the differences between the outsourcing risk scores regarding the type of IT activity considered for outsourcing: Application Management or Application Development. Application Management deals with existing applications, whereas Application Development considers applications more or less 'from scratch'.

Fig. 3 shows that there are no large differences between the different IT functions. Although the Application Complexity should not differ much between AD and AM (new applications can be built as complex as existing older ones) Organization Capability could be higher for Application Management. Reason for that is that for existing applications there was more time available to improve outsourcing risk factors like knowledge management, testing experiences, training, etc. When we take the outsourcing risk factors of Organization Capability for AM into consideration, we see in many cases that although (or is it maybe because of) the application is a couple of years old a lot of those factors are not that well in place.

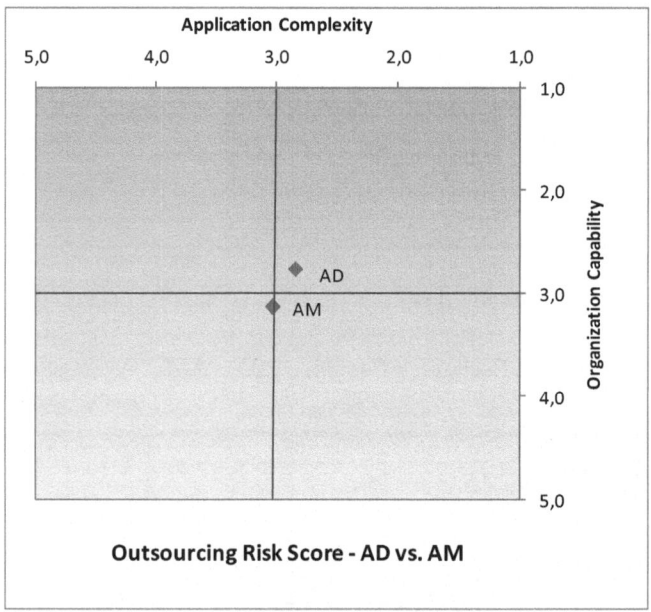

Fig. 3. Risk overview matrix – cross section AD / AM function

Package Based or Custom Made
Another interesting cross section is related to the type of application: package based or custom made. Every application based on a standard packaged application (e.g., SAP) is considered as package based, regardless the level of customization. One would expect that it is easier to outsource package based applications then custom made ones because package based applications are more standardized and built with more experiences in practice. For that reason we even defined one question about the

type of application where the risk score for the out-of –the-box implementation of a package based solution is low and custom made applications is high. Although we do not make a difference in this overview between out-of-the-box implementations or customized implementation of packages, this overview can be explained by another trend as well. Because packages are most of the time used for core applications, there is also a large impact on risk factors like the amount of business logic, interfacing, interaction with end users, and so on.

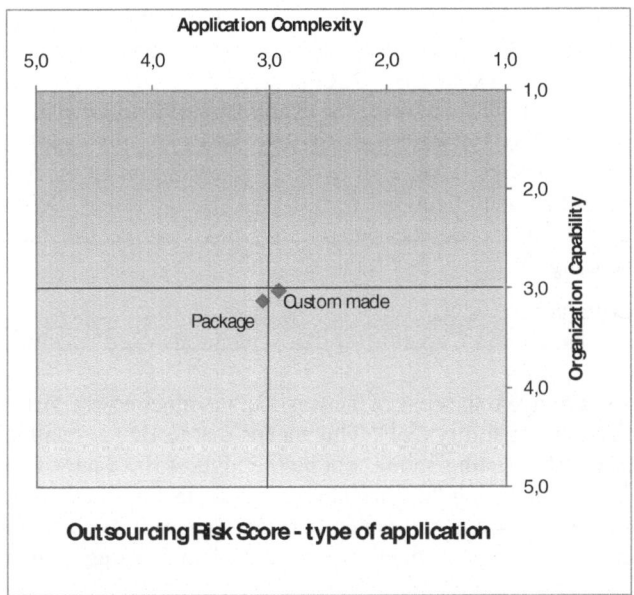

Fig. 4. Risk overview matrix – cross section Package based / Custom made

Table 4. Overview numbers per cross section

Type of application	AD	AM	Average complexity	Average capability
Package	6	15	2.9	3.0
Custom	1	27	3.2	3.2

3.3 Specific Findings of the Risk Factors

When we drill down to the risk factors per pillar we can also see interesting effects. First we will look at the risk factors in the Application Complexity pillar and after that we will look at Organization Capability.

Application Complexity
First fact, looking at all the data of the Application Complexity factors, is that the infra-structure risk factor was scored significantly lower (average score of 2.2) than the other factors and also lower than the average Application Complexity score of 3.1. Besides, the deviation between the individual infrastructure risk scores of the applications was

low. Looking at the specific cases, one can see that almost every application was running on only one or two different platforms with a limited amount of servers and that there was no entanglement of the infrastructure support and the AD or AM activities. From these facts we can conclude that the infrastructure is no real risk introducing factor for outsourcing related to the other factors of Application Complexity.

Second interesting effect is that the risk factors 'Architecture', 'Business Logic' and 'Interfacing' (average of respectively of 3.4, 3.3 and 3.3) have the highest impact on the average score of Application Complexity. So, given our RAS approach, those risk factors are the most important risk contributors in outsourcing an application.

The last effect we could see from our data set was the deviation between individual IT applications for the risk factors 'Architecture', 'Interfacing', and 'Interaction'. (std.dev. of 0.7 or more). There were, for example, applications with a simple architecture (an out-of-the-box software package like SAP) and also applications with all kinds of structures that were developed over the years. Some applications with a couple of interfaces, but also applications with over a 100 interfaces. So this might mean that if one wants to influence the outsourcing risk factors, the most changes are caused by influencing the above risk factors.

Organization Capability

Looking at the Organization Capability pillar we can also see two interesting effects over all cases.

First is that the average risk score of Testing (2.7) is much lower than the overall risk score of Organization Capability (3.1). This means that testing is relatively well organized in all cases and that testing should not be the highest risk factor for outsourcing in general. This is an interesting fact because the risk factor 'methodologies & tools' shows that AD & AM processes in general are not that well in place and also the risk factor ′Requirement Mgt′ scores much worse (3.2) than ′Testing′. The other four outsourcing risk factors have more or less the same average score.

The second effect is related to the great differences in score of the risk factors. In particular the risk factors ′Knowledge Management′ and ′Methodologies & Tools′ have a great difference in the scores between the cases (std. dev of 0.7). The cases with low scores on those two factors are for relatively many cases related to applications in the building phase (AD function). But also the other risk factors have a difference in scores with a standard deviation of 0.6, so from the Organizational Capability side you cannot pinpoint specific outsourcing risk factors which are less changeable.

4 Discussion

In this section we focus on the approach of measuring the outsourcing risks and explain to what extent the approach does not affect the results of the scores and therefore the findings in the last section. The three most important subjects in the approach are the data collection, the risk calculation and the setup of the outsourcing risk model.

Data collection

Key issue of data collection is the reliability of the data. The data filled in in the questionnaire is collected by IT consultants which might have had reasons to interpret the data in a specific direction. Of course they had to address the sources of the data and a team discussed the overall results, which minimized individual interpretations. But because not all questions are stated in an objective way, there is always a probability of an unacceptable level of subjectivity. When the questionnaire is improved by only addressing objective questions the reliability will be much better. Besides you can define a formal process with quality checks and training, e.g. like CMM assessments, so the reliability of the data collection itself will improve as well.

Another key question is to what extent the IT applications are chosen randomly. We have used our data from 10 organizations that were looking for the right sourcing scenario. Although those 10 organizations are randomly chosen, we cannot say for sure that all the 50 IT applications are chosen randomly related to all possible IT applications in the world. All the selected IT applications were subject to outsourcing and that might mean that they have some specific characteristics in themselves. The only way to find that out, is to measure outsourcing risks on a set of applications that are not subject to outsourcing. If the same kind of results are found as for IT applications subject to outsourcing, there is a proof that data reliability is acceptable.

Risk calculation
The basis of the approach is calculating two sum scores. This might affect the results negatively if the cases do not fulfill some preconditions. One of the effects is that the risk factor with the highest average score influences the average score per pillar mostly. When looking at the scores of all risk factors from both pillars, one can see that the differences in scores is not extreme (maximum difference of 1.1 and 0.6) and therefore the effect is not that large.

The second negative effect might be the negative correlation between some of the risk factors, which causes a suppression of the average score. In our correlation matrix we could not find any negative correlation between any risk factors. Given these checks we can conclude that there are no negative effects on our results.

Correlation between risk factors
What we do see, is the correlation between the two pillars Application Complexity and Organization Capability. That is not surprisingly because one can imagine that the maturity of the organization will affect some of the complexity elements of the application. E.g. if testing is not done well, there will be more changes to be made in the application or with inaccurate documentation managing an up-to-date architecture is a more difficult task. There are also different correlations between the risk factors within Application Complexity and Organization Capability.

Starting with Application Complexity we observe correlations between the risk factors ´Business Logic´, ´Application´ and ´Interfacing´. Also the risk factors ´Architecture´, ´Interaction´ and ´Support´ are correlated with each other. This means on the one hand that elements that affect the outsourcing risk factor are related to each other and if you can reduce the risk score of one it will positively affect other elements as well. On the other hand the average score of Application Complexity will be influenced strongly by those risk factors which are correlated.

The risk factor ´Infrastructure´ does not correlate with any other risk factor within Application Complexity. This means the average risk score of ´Infrastructure´ does not have a high impact on the average score of Application Complexity. This increases the effect of the low average score of ´Infrastructure´ we have seen in the section above . ´Infrastructure´ is not a high risk factor when we are deciding about outsourcing.

Looking at the Organization Capability risk factors we see a strong correlation between the first three risk factors (´Knowledge mgt.´, ´Methodologies & Tools´, ´Requirement Mgt.´) . The risk factor ´Testing´ is less correlated to the others and the ´people´ risk factor is not correlated at all. About the risk factors which are correlated to each other we can state the same conclusions as to the correlated factors within Application Complexity.

Setup of the model

In our outsourcing risk model we have used two pillars with several risk factors per pillar. The final result of the outsourcing risk measurement is the risk overview matrix, which has been called powerful by many IT managers. IT Managers recognize themselves in the risk overview matrix and give them the possibility to discuss with IT consultants about outsourcing scenarios given their situation. Also the two pillars are seen as good baselines for deciding about outsourcing, given the literature review by Jager [5]. Less straightforward is the choice of the outsourcing risk factors per pillar. Are those risk factors the only factors, are they positioned in the right pillar and to what extent do they contribute to the outsourcing risk level? Key to these remarks is the fact that which questions are underneath each factor and so which fact is important to a specific risk factor. If it is possible to define objective facts which influence the outsourcing risk and to what extent, see for example the discussion about ´Infrastructure´, it probably will strengthen the outsourcing risk model in both risk calculation and data collection.

5   Conclusions and Further Research

RAS is a methodology that is supporting IT managers and IT consultants quite well during the outsourcing decision making of IT applications. One of the powerful elements of RAS is the outsourcing risk measurement scores and the way the results are quantified and visualized in the risk overview matrix. In further research we like to investigate the effect of the outsourcing risks scores on the advised sourcing scenarios to IT managers. For instance, is Organization Capability indeed more leading than Application Complexity when defining a sourcing scenario as stated in [5]?

Given the 50 IT applications, we have addressed the following findings related to outsourcing risk measurements. Related to the IT application itself the outsourcing risk factors ´Architecture´, ´Business Logic´ and ´Interfacing´ have a higher impact than other factors. The risk factor ´Infrastructure´ has a low impact on the outsourcing risks. Looking at the capability of the organizations we have seen that ´Testing´, as one of the outsourcing risk factors, is not a high risk factor. Unfortunately we did not see any trends yet in cross sections like AD vs. AM or Package Based vs. Custom made applications, even though one might expect there would be differences. So in general we can state that some outsourcing risk factors have a higher impact on outsourcing than others. There are also specific Application Complexity risk factors which are less changeable than others, so they are standard´ risks for every application to be outsourced.

Apparently there are some remarks about the approach of measuring the outsourcing risks. The approach is based on quantifying the outsourcing risks by calculating sum scores. We have checked some basic mathematical characteristics and we could not find any inadequacies. Yet, improvements can be made by detaching some risk factors from each other, so correlation of risk factors does not influence the overall score that much. Also the reliability of data can be increased by looking for objective characteristics of the IT application and the organization, so the data collection will be less depended of the knowledge and experience of the assessor.

Given all these conclusions we can state that given this approach we showed some interesting findings, which of course give input to new research questions. Improving the outsourcing risk measurement model will generate much more data for investigations. Besides, it will probably give better outsourcing advise to organizations – to prove that, first an analysis should take place on the outsourcing risks scores on the one hand and the advice given about the sourcing scenario on the other.

References

1. Beulen, E.: Global Sourcing, Inaugural Speech, Tilburg University (2008)
2. Cohen, L., Young, A.: Multisourcing. Harvard Business School Press, Boston (2006)
3. Capgemini, Global CIO Survey 2007 – IT Agility (2007)
4. EquaTerra, Outsourcing Performance (in Dutch) (2008)
5. Jager, C.J., Vos, S., Borgers, M., Harmsen, F., Brinkkemper, S., Wijngaert, L.: Controlling risk prior to offshore application development. In: Proceedings Global Sourcing Workshop, Val d'Isère (2008), http://www.globalsourcing.org.uk/papers.htm
6. Schoeman, S., Bakker, N., Borgers, M., Van Hillegersberg, J., Moody, D.: Bridging the Gap Between the Theory and Practice of IT Outsourcing Strategy Design. In: Proceedings Global Sourcing Workshop, Val d'Isère (2008),
   http://www.globalsourcing.org.uk/papers.htm
7. Steenbeek, W., van den Wijngaert, L., Harmsen, F., Brinkkemper, S., van den Brand, M.: Sourcing decision-making: Eliciting consultancy knowledge using Policy Capturing. In: Proceedings European Conference on Information System, Regensburg (2005)
8. Willcocks, L.P., Lacity, M.C.: Global sourcing of business & IT services, Palgrave McMillan (2006)

Author Index